DOTS AND LINES

DOTS
AND
LINES

Richard J. Trudeau

THE KENT STATE UNIVERSITY PRESS

Copyright © 1976 by the Kent State University Press
All rights reserved
ISBN 0-87338 (cloth)
    0-87338 (paper)
Library of Congress Catalog Card Number 78-16894
Manufactured in the United States of America

Library of Congress Cataloging in Publication Data

Trudeau, Richard J
    Dots and lines.

  Includes bibliographies and index.
  1. Graph theory. I. Title.
QA166.T74 1978   512'.22   78-16894

To Dick Barbieri and Chet Raymo

TABLE OF CONTENTS

PREFACE

This book is about pure mathematics in general, and the theory of graphs in particular. ("Graphs" are networks of dots and lines; they have nothing to do with "graphs of equations.") I have interwoven the two topics, the idea being that the graph theory will illustrate what I have to say about the nature and spirit of pure mathematics, and at the same time the running commentary about pure mathematics will clarify what we do in graph theory.

I have three types of reader in mind.

First, and closest to my heart, the mathematically traumatized. If you are such a person, if you had or are having a rough time with mathematics in school, if you feel mathematically stupid but wish you didn't, if you feel there must be *something* to mathematics if only you knew what it was, then there's a good chance you'll find this book helpful. It presents mathematics under a different aspect. For one thing, it deals with *pure* mathematics, whereas school mathematics (geometry excepted) is mostly *applied* mathematics. For another, it is a more qualitative than quantitative study, so there are few calculations.

Second, the mathematical hobbyist. I think graph theory makes for marvelous recreational mathematics; it is intuitively accessible and rich in unsolved problems.

Third, the serious student of mathematics. Graph theory is the oldest and most geometric branch of topology, making it a natural supplement to either a geometry or topology course. And due to its wide applicability, it is currently quite fashionable.

The book uses some algebra. If you've had a year or so of high school algebra that should be enough. Remembering specifics is not so important as having a general familiarity with equations and inequalities. Also, the discussion in Chapter 1 presupposes some experience with plane geometry. Again no specific knowledge is required, just a feeling for how the game is played.

Chapter 7 is intended for the more mathematically sophisticated reader. It generalizes Chapters 3–6. It is more conceptually difficult and concisely written than the other chapters. It is not, however, a prerequisite for Chapter 8.

The exercises range from trivial to challenging. They are not arranged in order of difficulty, nor have I given any other clue to their difficulty, on the theory that it is worthwhile to examine them all.

The suggested readings are nontechnical. Those that have been starred are available in paperback.

There are a number of more advanced books on graph theory, but I especially recommend *Graph Theory* by Frank Harary (Addison-Wesley, 1969). It contains a wealth of material. Also, graph theory's terminology is still in flux and I have modeled mine more or less after Harary's.

Richard J. Trudeau
July 1975

1. PURE MATHEMATICS

Introduction

This book is an attempt to explain pure mathematics. In this chapter we'll talk about it. In Chapters 2–8 we'll *do* it.

Most pre-college mathematics courses are oriented toward solving "practical" problems, problems like these:

> A train leaves Philadelphia for New York at 3:00 PM and travels at 60 mph. Another train leaves New York for Philadelphia at 3:30 PM and travels at 75 mph. If the distance between the cities is 90 miles, when and at what point will the trains pass?

> If a 12-foot ladder leaning against a house makes a 75° angle with the ground, how far is the foot of the ladder from the house and how far is the top of the ladder from the ground?

Mathematics that is developed with an eye to practical applications is called "applied mathematics". With the possible exception of Euclidean geometry, pre-college mathematics is usually applied mathematics.

There is another kind of mathematics, called "pure mathematics", which is a charming little pastime from which some people derive tremendous enjoyment. It is also the basis for applied mathematics, the "mathematics" part of applied mathematics. Pure mathematics is *real* mathematics.

To understand what mathematics is, you need to understand what pure mathematics is. Unfortunately, most people have either seen no pure mathematics at all, or so little that they have no real feeling for it. Consequently most people don't really understand mathematics; I think this is why so many people are afraid of mathematics and quick to proclaim themselves mathematically stupid.

Of course, since pure mathematics is the foundation of applied mathematics, you can see the pure mathematics beneath the applications if you look hard enough. But what people see, and remember, is a matter of emphasis. People are told about bridges and missiles and computers. Usually they don't hear about the fascinating intellectual game that lies beneath it all.

Earlier I implied that Euclidean geometry—high school geometry— might be an example of pure mathematics. Whether it is or not again depends on emphasis.

Euclidean geometry as pure mathematics

What we call "Euclidean geometry" was developed in Greece between 600 and 300 B.C., and codified at the end of that period by Euclid in *The Elements*. *The Elements* is the archetype of pure mathematics, and a paradigm that mathematicians have emulated ever since its appearance. It begins abruptly with a list of definitions, followed by a list of basic assumptions or "axioms" (Euclid states ten axioms, but there are others he didn't write down). Thereafter the work consists of a single deductive chain of 465 theorems, including not only much of what was known at that time of geometry, but algebra and number theory as well. Though that's quite a lot for one book, people who read *The Elements* for the first time often get a feeling that things are missing: it has no preface or introduction, no statement of objectives, and it offers no motivation or commentary. Most strikingly, there is no mention of the scientific and technological uses to which many of the theorems can be put, nor any warning that large sections of the work have no practical use at all. Euclid was certainly aware of applications, but for him they were not an issue. To Euclid a theorem was significant, or not, in and of itself; it did not become more significant if applications were discovered, or less so if none were discovered. He saw applications as external factors having no bearing on a theorem's inherent quality. The theorems are included *for their own sake*, because they are interesting in themselves. This attitude of self-sufficiency is the hallmark of pure mathematics.

The Elements is the most successful textbook ever written. It has gone through more than a thousand editions and is still used in some parts of the world, though in this country it was retired around the middle of the nineteenth century. It is amazing that it was used as a school text at all, let alone for 2200 years, as it was written for adults and isn't all that easy to learn from.

Of the modern texts that have replaced *The Elements,* many are faithful to its spirit and present geometry as pure mathematics, *ars gratia artis.* There are others, however, that have sandwiched around Euclid's work long discussions of the "relevance" and "practicality" of geometry, complete with pictures of office buildings and space capsules and the suggestion that geometry is primarily a branch of engineering. Of course, applied geometry is a perfectly valid science, but I can't help objecting to such books on the ground that they rob school children of their only encounter with pure mathematics.

Despite this problem, in the next section I shall use Euclidean geometry as an example of pure mathematics, as it is the branch of pure mathematics with which you are most likely to be familiar.

Games

Basically pure mathematics is a box of games. At last count it contained more than eighty of them. One of them is called "Euclidean geometry". In this section I will compare Euclidean geometry to chess, but you won't have to be a chessplayer to follow the discussion.

Games have four components: objects to play with, an opening arrangement, rules, and a goal. In chess the objects are a chessboard and chessmen. The opening arrangement is the arrangement of the pieces on the board at the start of the game. The rules of chess tell how the pieces move; that is, they specify how new arrangements can be created from the opening arrangement. The goal is called "checkmate" and can be described as an arrangement having certain desirable properties, a "nice" arrangement.

In Euclidean geometry the *objects* are a plane, some points, and some lines. The plane corresponds to a chessboard, the points and lines to chessmen. The *opening arrangement* is the list of axioms, which are accepted without proof. The analogy with the opening arrangement of chessmen may not be apparent, but it is quite strong. First, the opening arrangement of chessmen is *given;* to play chess you must start with that arrangement and no other. In the same way the axioms of Euclidean geometry are given. Second, the opening arrangement of chessmen specifies how the objects with which the game is played are related at the outset. This is exactly what the axioms do for the game of Euclidean geometry; they tell us, for example, that points and lines lie in the plane, that through two points there

passes one and only one line, etc. The *rules* of Euclidean geometry
are the rules of formal logic, which is nothing but an etherealized
version of the "common sense" we absorb from the culture as we
grow up. Its rules tell us how statements can be combined to produce
other statements. They tell us, for example, that the statements "All
men are mortal" and "Plato is a man" yield the statement "Plato
is mortal." (The example is Aristotle's. He wrote the first book on
logic by recording patterns of inference he saw people using every
day.) In particular the rules of logic tell us how to create, from the
opening arrangement (the list of axioms), new arrangements (called
"theorems"). And the *goal* of Euclidean geometry is to produce as
many "nice" arrangements as possible, that is to prove profound and
surprising theorems. Checkmate terminates a chessmatch, but Euclidean geometry is open-ended.

Games have one more feature in common with pure mathematics.
It is subtle but important. It is that the objects with which a game
is played have no meaning outside the context of the game.

Chessmen, for example, are significant only in reference to chess.
They have no necessary correspondence with anything external to
the game. Of course, we could *interpret* the pieces as regiments at
the First Battle of the Marne, and the board as French countryside.
Or an interpretation could be brought about by a wager, say each
piece represents $5.00 and losing it means paying that amount. But
no such correspondence between the game and things outside the
game is necessary.

You may balk at this, since, for historical reasons, chessmen have
names and shapes that imply an essential correspondence with the
external world. The key word is "essential"; there is indeed a correspondence, but it is inessential. After all, chessmen require names
of some sort, and must be shaped differently to avoid confusion, so
why not call them "kings", "queens", "bishops", "knights", etc., shape
them accordingly, and trade on the image of excitement and competition
thereby created? Doing so is harmless and makes the game more
popular. But this particular interpretation of the game, like all others,
has nothing to do with chess *per se*.

Here's an example. Suppose we substitute silver dollars for kings,
half-dollars for queens, quarters for bishops, dimes for knights, nickels
for rooks, pennies for pawns, and an eight-by-eight array of chartreuse
and violet circles for the standard board, but otherwise follow the
rules of chess. Such a game would look strange and even sound
strange—"pawn to king four" would now be "penny to silver dollar
four"—but surely if we were to play this apparently unfamiliar game,

there would be no doubt that we are playing the familiar game of chess. Indeed, chessmasters sometimes play without a board or pieces of any kind; they merely announce the moves and keep track in their heads. Two such people are still playing chess, for after all they say they are, and they certainly should know.

It appears then that the essence of chess is its abstract structure. Names and shapes of pieces, colors of squares, whether the "squares" are in fact square, even the physical existence of board and pieces, are all irrelevant. What is relevant is the number and geometric arrangement of the "squares", the number of types of piece and the number of pieces of each type, the quantitative-geometric power of each piece, etc. Everything else is a visual aid or a fairy tale.

So it is with pure mathematics. Euclid's words "plane", "point", and "line" suggest that geometry deals with flat surfaces, tiny dots, and stretched strings, but this implied interpretation of geometry is only that. It is analogous to the interpretation of chess as a battle. Geometry is no more a study of flat surfaces and dots than chess is a military exercise. As in any game, the objects geometers play with, and consequently their arrangements—the axioms and theorems—have no necessary correspondence with things external to the game.

In support of this let me point out that geometers never define the words "plane", "point", or "line". (Euclid offered an intuitive explanation but did not actually define them; moderns leave the words undefined.) So *no one knows* what planes, points, or lines are, except to say that they are objects which are related to one another in accordance with the axioms. The three words are merely convenient names for the three types of object geometers play with. Any other names would do as well. Were we to attack Euclid's *Elements* with an eraser and remove every occurrence of the words "plane", "point", and "line", replacing them respectively with the symbols "#", "$", and "?", the result would still be *The Elements* and the game would still be geometry. To a casual observer the vandalized *Elements* wouldn't look like geometry; what had been "two points determine one and only one line" would now be "two $'s determine one and only one ?". But then two people hunched over a board of chartreuse and violet circles, littered with coins, doesn't look like a chessmatch. The game would still be geometry because it would be structurally identical to geometry. And were we to further maim *The Elements* by erasing all the diagrams, it still wouldn't make a difference. Geometric diagrams are to geometers what board and pieces are to chessmasters: visual aids, helpful but not indispensable.

Why study pure mathematics?

There emerges from the foregoing an image of pure mathematics as a meaningless intellectual pastime. Yet carved over the door to Plato's Academy was the admonition, "Let no one ignorant of geometry enter here!" And pure mathematics has been held in the highest regard ever since. It would seem to have no more to recommend its inclusion in school curricula than, for instance, chess, yet it is universally favored by academics over other games. I shall give three reasons for this.

Pure mathematics is applicable. Because pure mathematics has no inherent correspondence with the outside world, we are free to make it correspond, to interpret it, in any way we choose. And it so happens—this is the interesting part—that most branches of pure mathematics can be interpreted in such a way that the axioms and theorems become approximately true statements about the external world. In fact, some branches have several such interpretations.

Pure mathematics that has been made to correspond in this way to the world outside is called "applied mathematics". Pure mathematics is Euclid saying "three \$'s not on the same ? determine a unique #." Applied mathematics is a surveyor reading Euclid, interpreting "\$", "?", and "#" in a way that seems in accord with the axioms, and concluding that a tripod would be the most stable support for his telescope.

On one level, the applicability of pure mathematics is no surprise. Just as chess (as we know it) has been modeled on certain aspects of medieval warfare, even though strictly speaking the game has nothing to do with warfare, so too most branches of pure mathematics have started as models of physical situations. A branch of pure mathematics utterly lacking in significant interpretations would be boring to the community of pure mathematicians and would soon die out from lack of interest. Though they are unconcerned with applications as such, pure mathematicians are like most people in that they find it hard to be enthusiastic about something unless, under some aspect at least, it has the spontaneous appearance of truth.

But on a deeper level, the applicability of pure mathematics is quite mysterious. It's true that pure mathematics often originates with an abstraction from the physical world, as geometry begins with idealized dots and strings and tabletops, but the tie is only historical. Once the abstractions have been made the mathematical game comes into independent existence and evolves under its own laws. It has no necessary correspondence with the original physical situation. The mathematician does not deal with physical objects themselves, but with idealizations that exist independently and differ from their physical counterparts in a great many respects. And entirely within his own mind, the mathematician subjects these abstractions to a reflective, self-analytic process, a process in which he is trying to learn about himself, to learn what in a sense he already knows. This process is strictly internal to a human mind—a Western mind at that—and so is presumably different from whatever the process by which the physical situation evolves; yet when the mathematician compares his results to outside events, he often finds that nature has evolved to a state remarkably like his mathematical model. That the universe is so constructed has seemed uncanny to many famous mathematicians and scientists, moving them to comment in a mystical fashion that seems totally out of character:

"Number rules the universe."

—Pythagoras

"Mathematics is the only true metaphysics."

—Lord Kelvin

"How can it be that mathematics, being after all a product of human thought independent of experience, is so admirably adapted to the objects of reality?"

—Einstein

"The Great Architect of the Universe now begins to appear as a pure mathematician."

—Sir James Jeans

(Quotations from E. T. Bell's *Men of Mathematics*.)

Applicability is the chief difference between the games known collectively as "pure mathematics" and other games. There's some kind of chemistry involving nature, and people, and pure mathematics, that enables applied mathematics to predict the future, whereas mankind has yet to make a success of applied chess.

Pure mathematics is a culture clue.

". . . common sense is, as a matter of fact, nothing more than layers of preconceived notions stored in our memories and emotions for the most part before age eighteen."

—Albert Einstein

Our common sense, or world view, is not "common" to all people. It is shaped by the culture we inhabit. It is like a pair of glasses few of us ever manage to take off, so of course we see confirmation everywhere we look.

Much of Western intellectual tradition has been inherited from the Greeks. Our science and philosophy in particular are shot through with beliefs and opinions and forms of speech that were once explicit doctrines of Plato, Aristotle, and the like, but have come to be embedded anonymously in the fabric of our thought. Of this embedded material perhaps the most fundamental is logic, the standard by which we judge reasoning to be "correct", a standard first written down by Aristotle in *The Organon* (about 350 B.C.).

Is logic itself "correct"? Some Eastern philosophers would call it "ignorance". I use logic all the time in mathematics, and it seems to yield "correct" results, but in mathematics "correct" by and large means "logical", so I'm back where I started. I can't defend logic because I can't remove my glasses.

"Correct" or not, logic is basic to Western rationality and to the whole scientific enterprise. And not surprisingly, since logic is the study of deduction and pure mathematics is the only completely deductive study, logic is inextricably intertwined with pure mathematics. I think this is the chief reason for the prominence of mathematics in our schools. Logic is a fundamental component of the culture, so the culture quite naturally sets a premium on teaching the next generation to think in logical categories.

Incidentally, there's a lot of debate on which came first, logic or mathematics. In one sense logic is prior to mathematics, as mathematics uses the laws of logic. But Aristotle abstracted the laws of logic at least in part from the pure mathematics he studied at Plato's Academy, so in another sense mathematics is more basic. G. Spencer Brown argues this position in *Laws of Form,* p. 102:

> A theorem is no more proved by logic and computation than a sonnet is written by grammar and rhetoric, or than a sonata is composed by harmony and counterpoint, or a picture painted by balance and perspective. Logic and computation, grammar and rhetoric, harmony and counterpoint, balance and perspective, can be seen in the work *after* it is created, but these forms are, in the final analysis, parasitic on, they have no existence apart from, the creativity of the work itself. Thus the relation of logic to mathematics is seen to be that of an applied science to its pure ground, and all applied science is seen as drawing sustenance from a process of creation with which it can combine to give structure, but which it cannot appropriate.

Pure mathematics is fun. At this moment there are thousands of people around the world doing pure mathematics. A few might be doing so because they foresee a possible application. A few might be philosophers taking Bertrand Russell's advice that "to create a healthy philosophy you should renounce metaphysics but be a good mathematician." There might even be a few ascetics who are doing it to sharpen their minds. But the vast majority are doing it simply because it's fun.

Pure mathematics is a first-rate intellectual adventure, ". . . an independent world/Created out of pure intelligence" (Wordsworth) that is neither science nor art but somehow partakes of both.

Pure mathematics is the world's best game. It is more absorbing than chess, more of a gamble than poker, and lasts longer than Monopoly. It's free. It can be played anywhere—Archimedes did it in a bathtub. It is dramatic, challenging, endless, and full of surprises.

Pure mathematics is a pleasant way to pass the time until the end. And to me that makes it very serious, very important indeed.

What's coming

Talking about pure mathematics isn't enough. To really understand what it is, you have to do it. That's why this book has seven more chapters.

In the pages ahead we shall develop the rudiments of one rather modest game from the pure mathematics game-box. It has the unfortunate name "graph theory". The name is unfortunate because it is misleading. The objects with which the game is played are called "graphs", which is also the name given to the pictures of equations drawn in high school algebra courses. "Graph" in our sense of the word is not a picture of an equation, but rather a network of dots and lines. "Network theory" would be a better name for our game. Nevertheless the official name of the game is "graph theory", and mathematicians just have to remember that the "graph" of "graph theory" is not an algebraic graph. You'll appreciate the difference after reading Chapter 2.

Pure mathematics is a *big* game-box. I selected graph theory because it has several features that seemed desirable for this kind of book:

Graph theory is new; the bulk of it has been developed since 1890.

Graphs are simple, intuitively accessible things. In a book of this size we can develop enough graph theory to prove several spectacular theorems, and to half-prove several others.

Graph theory's frontier is easily reached, and you will find yourself, probably for the first time, thinking about famous unsolved problems.

If, like so many people, you dislike mathematics, it may be that you merely dislike applied mathematics, in which case you'll enjoy working your way through the chapters ahead. But mathematics books don't read like novels, so take your time.

If, on the other hand, you like mathematics, it may be that you like only applied mathematics, in which case you won't enjoy this book at all. Though graph theory has many applications (to electrical circuitry, chemistry, industrial management, linear programming, game theory, transportation networks, statistical mechanics, social psychology, and more) we're going to ignore them. After all, the book is supposed to be an antidote for an overdose of applied mathematics.

By this book I hope to influence your present attitude toward mathematics. Many people have a distorted notion of the nature and spirit of mathematics. Whether after reading the book you like or dislike mathematics is beside the point. My goal is rather that you form your attitude toward the subject in response to a clear image of it, not a phantasm.

Suggested reading

A Mathematician's Apology by G. H. Hardy (Cambridge University Press, 1967). A once-great has-been's defense of his life's work. I highly recommend it.

*Part II of *Fantasia Mathematica,* edited by Clifton Fadiman (Simon & Schuster, 1958). Part II is a collection of science-fiction stories with mathematical premises.

Mathematics in Western Culture by Morris Kline (Oxford University Press, 1964). Kline argues that mathematics has been a major cultural force in the West, influencing not only science and philosophy, but religion, politics, painting, music, and literature as well.

The Nature and Growth of Modern Mathematics (2 vols.) by Edna E. Kramer (Fawcett, 1970).

* *Philosophy of Mathematics* by Stephen F. Barker (Prentice-Hall, 1964).

* *Laws of Form* by G. Spencer Brown (Bantam, 1973). Heavy going; brilliant.

2. GRAPHS

Introduction

The things in Figure 1 are called "graphs", and are typical of what we will be playing with. Despite the name, they are unrelated to the pictures of equations drawn in high school algebra courses. To quickly convince you of this, I mention that we will consider graphs a), b), and c) to be identical (the word we will use is "isomorphic"), though if drawn in cartesian coordinates they would depict three different equations. In order to define "graph" more precisely, we must first discuss "sets".

Sets

Definition 1. A *set* is a collection of distinct objects, none of which is the set itself.

If you've encountered sets before you may find the last part of the definition a bit puzzling. We'll return to it in the next section, but for now suffice it to say that we intend to exclude collections like $A = \{1, 2, 3, A\}$.

The words "collection" and "object" are left undefined. Strictly speaking, therefore, we don't know what we're talking about. That's the way it is with pure mathematics. But there is implied an interpretation of "set" as being a bunch of things, an aggregate of entities, etc., and if your intuition feels more comfortable with something to hold onto (mine does), feel free to think of a set as being just what it sounds like. But be careful, for the words "collection" and "object" are used in a slightly unconventional fashion. We are allowing "collections" of infinitely many things, or just one thing, or even

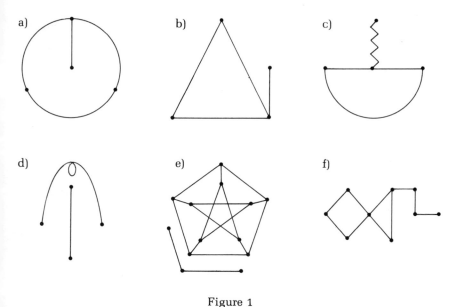

Figure 1

no things. And an "object" need not be something we can smell or trip over; to a mathematician an "object" is anything conceivable, including numbers, unicorns, and Peter Pan. The only things left out are the inconceivable, for instance a figure which simultaneously has all the geometric properties of both a triangle and a circle. Everything else is acceptable.

A set is usually denoted by listing the names of the objects, separated by commas, within a pair of braces. So

{12, q, $, Empire State Building, 9, King Arthur}

is a set. Notice that the objects, which are called the *elements* of the set, need not have anything in common; mathematical sets are not like sets of luggage. Usually we give sets names to avoid a lot of writing when we subsequently refer to them. Thus one might say, "Let A = {12, q, $, Empire State Building, 9, King Arthur}." This is the beginning of mathematical notation, something which, like any other shorthand, can be more confusing than helpful if not learned gradually as it comes along. Definitions are just another version of the same thing. I recommend that you memorize (yes!) the meanings of all new terms and symbols when they first appear.

Definition 2. A set containing no elements is called a *null set* or an *empty set*.

An empty set plays the same role in set theory that 0 plays in arithmetic, but you must be careful not to confuse the two. An empty set is a set; 0 is a number. They are of completely different species. Think of it this way: 0 is a number that tells you how many objects there are in an empty set.

Definition 3. A set A is said to be a *subset* of a set B, denoted "$A \subset B$", if every element of A is also an element of B.

Example. If $A = \{1, 2, 3\}$ and $B = \{\&, 3, +, 1, 2\}$, then $A \subset B$, $A \subset A$, and $B \subset B$. Notice that every set is a subset of itself.

Convention. We agree to consider an empty set to be a subset of every set.

This "convention" can in fact be proved, but only by a logical contortion that seems inappropriate at the present. Thus it is a "convention" instead of a "theorem".

Example. If J is an empty set and A and B are the sets of the previous example, then $J \subset A$, $J \subset B$, and $J \subset J$.

Definition 4. A set A is said to be *equal* to a set B, denoted "$A = B$", if $A \subset B$ and $B \subset A$.

Thus $A = B$ if A and B consist of exactly the same objects. It follows that the order in which the elements of a set appear is irrelevant. For example, $\{1, 2, 3\} = \{2, 1, 3\}$ since $\{1, 2, 3\} \subset \{2, 1, 3\}$ and $\{2, 1, 3\} \subset \{1, 2, 3\}$.

Theorem 1. *There is only one empty set.*

Proof. Let J and K be empty sets. Then by the Convention, $J \subset K$ and $K \subset J$, so $J = K$ by Definition 4. Thus all empty sets are equal; that is, there is really only one empty set.

Notation. The empty set (we had to say "an empty set" before, but now we can say "the empty set") shall be denoted by "$\emptyset$" or "$\{ \ \}$".

Use either notation at your whim, but be careful not to combine them. "$\{\emptyset\}$" would be interpreted by a mathematician as signifying a set containing one element, that element being the symbol $\emptyset$.

In view of the word "distinct" in Definition 1, collections like $\{1, 3, 3, \&, 407\}$, in which an object is listed more than once, are not sets. The definition was designed to exclude this sort of thing because

"{1, 3, 3, &, 407}" is a redundancy. We are being told twice that the collection contains the number 3.

All the set theory we'll need to talk about graphs has now been developed. But before going on I want to take time out and explain why in pure mathematics we are so excruciatingly careful about definitions and proofs (something you may have already noticed). In the process I'll explain the reason for that last clause in Definition 1. You may find parts of the next section a little involved, but I think my point is important and I encourage you to persist.

Paradox

The story begins in the sixth century B.C. with the Pythagoreans, a community of men and women founded by Pythagoras at Crotona in southern Italy. The Pythagoreans shared quasi-religious rituals, dietary laws, and devotion to mathematics as the key to understanding nature. It is they who first discovered that intuition and logic can disagree.

The Pythagorean Paradox. It happened like this. Let AB and CD be two finite straight lines. We will say that a finite straight line XY is a "common measure" of AB and CD if there are whole numbers m and n so that XY laid down m times is the same length as AB, and XY laid down n times is the same length as CD. For example, if AB were a yard long and CD 10 inches, a line segment XY of 2 inches would be a common measure with $m = 18$ and $n = 5$; for laying down XY eighteen times would produce a length of 36 inches, the same as AB, and laying down XY five times would produce a length of 10 inches, the same as CD. $XY = 1/4$ inch would be another common measure, this time with $m = 144$ and $n = 40$. It was intuitively evident to the Pythagoreans, and as I write it is intuitively evident to me, that a common measure can be found for *any* pair of finite straight lines—though of course it may be necessary to take XY quite small in order to measure both AB and CD exactly. Since $AB/CD = m(XY)/n(XY) = m/n$, a "rational" number (that is, a quotient of whole numbers), what my intuition predicts is that the quotient of two lengths is always a rational number.

Now take a square with side equal to 1 and draw a diagonal. By the Pythagorean Theorem the length of the diagonal is $\sqrt{2}$ and so the quotient of the length of the diagonal and the length of one of the sides is also $\sqrt{2}$. If the Pythagoreans' intuition and mine are correct

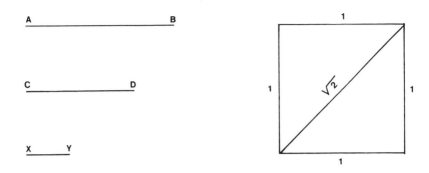

Figure 2

in asserting that the quotient of two lengths is always a rational number, $\sqrt{2}$ must be a rational number. But one of the later Pythagoreans (probably Hippasus of the fifth century B.C.) discovered, by an argument not based primarily on intuition, that $\sqrt{2}$ cannot be a rational number.

The proof went something like this. Any rational number can be reduced to "lowest terms", that is, expressed as a quotient of whole numbers that have no factors (other than 1) in common; for example $360/75 = 24/5$, and 24 and 5 have no common factor. Therefore, if $\sqrt{2}$ were a rational number, it would be possible to express $\sqrt{2}$ as $\sqrt{2} = p/q$ where p and q are whole numbers with no common factor. Squaring both sides gives $2 = p^2/q^2$, and multiplying both sides by q^2 gives $2q^2 = p^2$. This means that p^2 must be even, because it is twice another whole number, q^2. The Pythagoreans had previously proved that the square of an odd number is odd, so the fact that p^2 is even implies that p is even also (if p were odd, p^2 would be odd). So p is even; that is, p is the double of some other whole number; in symbols, $p = 2r$ for some whole number r. Substituting $p = 2r$ into the equation $2q^2 = p^2$ gives $2q^2 = (2r)^2$ or $2q^2 = 4r^2$. Dividing both sides by 2 gives $q^2 = 2r^2$, so q^2, being twice the whole number r^2, is even, and for the same reason as before q is even too. But if p and q are both even, as we have shown, then they have a common factor of 2, which contradicts our earlier statement that p and q have no common factor. Hence the assumption that $\sqrt{2}$ is rational leads to a contradiction and so is logically untenable.

At this point the Pythagoreans were at an impasse, what mathematicians call a "paradox". They were *sure*, on intuitive grounds, that $\sqrt{2}$, being a quotient of two lengths, is a rational number. On the other hand they were equally sure, on grounds of logic and computation, that $\sqrt{2}$ is not a rational number. Had they decided to accept intuition as more reliable than logic, the future of mathematics would have been quite different; but of course they decided in favor of logic, and mathematicians ever since have been taught to mistrust intuition. (I think this has something to do with the stereotype that mathematicians are "cold" people.)

To say that mathematicians consider intuition unreliable, however, does not mean that they have banished it from mathematics. On the contrary, the basic assumptions from which any branch of pure mathematics proceeds—the axioms—are unsusceptible to proof, and are accepted primarily because of intuitive appeal. And intuition plays a big role in the discovery of theorems as well, or mathematicians would be spending most of their time trying to prove false theorems.

It's just that intuitive evidence is not accepted as conclusive. Definitions and axioms are carefully framed with an eye to hidden assumptions. Time and again in the history of mathematics, paradoxes have arisen because fundamental notions were not divorced from their origins in the physical world. (The Pythagorean paradox is a good example: any carpenter can tell you that given two sticks AB and CD, it is always possible to cut a stick XY that will be a common measure; but mathematical straight lines *aren't* sticks.) Similarly, theorems are stated as precisely as possible; and though they are discovered by intuition, they are demonstrated by logic alone. This habit of mind is called by mathematicians "rigor". It is characteristic of pure mathematics.

In recent years pure mathematics has become "rigorous" beyond the Greeks' wildest dreams. The reason for this is an insidious paradox that surfaced around the turn of the century.

Russell's Paradox. During the last half of the nineteenth century several branches of pure mathematics were being simultaneously rebuilt because of paradoxes that had arisen a few decades earlier. This rebuilding was based on the notion of "set", which was defined as follows:

Old Definition. A set is a collection of distinct objects.

The set concept seemed like a good foundation because it was simple and obviously free of logical difficulty. Mathematicians were confident that with it they could build a logical edifice so solid as to never again be shaken by a paradox.

Then, in 1902, Bertrand Russell presented his famous paradox. He observed that if we define a set to be merely a collection of distinct objects, we have to include self-referent collections like $A = \{1, 2, 3, A\}$, in which the set is one of its own elements. Of course, the sets mathematicians would normally have occasion to discuss are not of this type, so he called these new sets "extraordinary". Standard sets like $B = \{3, 17, \&, \#, \text{Massachusetts}\}$, in which the name of the set does not appear on the list of elements, he called "ordinary". So an extraordinary set is a set that is an element of itself, and an ordinary set is a set that is not an element of itself.

Don't confuse "being an element of" and "being a subset of". Every set is a subset of itself, but only a strange set like $A = \{1, 2, 3, A\}$ is an element of itself.

The paradox unfolds by letting S be the collection of all ordinary sets and nothing else. I claim that S is a set. We are using the old definition that says a set is merely a collection of distinct objects, an "object" being anything conceivable. Ordinary sets are certainly conceivable (I am holding the concept in my mind as I write), so they are objects; S is a collection of such objects, so S is a set.

Every set is either ordinary or extraordinary, so S must be either ordinary or extraordinary. *The rub is that S is neither.* Recall what we know about S: it is a set, it contains every ordinary set, and it contains nothing else. If S itself were an ordinary set, it would be an element of itself and therefore extraordinary. On the other hand, if S were an extraordinary set, it would not be an element of itself (as it contains only ordinary sets), and therefore S would be ordinary. So S can be neither ordinary nor extraordinary, though being a set it must be one or the other. We are caught in a bewildering paradox.

One way out would be to abandon the so-called "Law of Excluded Middle", a rule of logic that says "any meaningful statement is either true or false and there is no other possibility." We used the Law of Excluded Middle in the above argument when we insisted that every set is either ordinary or not ordinary (i.e., extraordinary). If there were a "middle" possibility, the argument that S is neither ordinary nor extraordinary would lose its devastating effect. You may feel that the Law of Excluded Middle is so basic to reason that rejecting it is unthinkable, but there is a minority school of mathematicians, the Intuitionists, who do precisely that. The majority of mathematicians, however, have confidence in this part of Aristotle's logic, so when the paradox was announced they searched for another way out.

In 1910 Russell resolved his own paradox with what he called the "theory of types", which essentially says that we can avoid the paradox if we tack an extra clause onto the definition of "set" excluding collections that are elements of themselves. This is what we did in

Definition 1. Under this new definition there are no such things as "extraordinary" sets; all sets are ordinary, so S becomes the collection of all sets. Further, S itself is no longer a set, for if S were a set it would be an element of itself. S is in some sense "too big" to be a set. For collections like S we can invent a new name, say "class", to indicate that they are qualitatively different from sets. So the paradox is gone, leaving the following debris: S is not an extraordinary set because extraordinary sets don't exist; neither is S an ordinary set because S is not a set at all; S is a class.

The subtlety of Russell's paradox can be seen in the fact that mathematicians all over the world had worked with the theory of sets for more than twenty years before anyone noticed the big loophole in the very definition of "set". Standards of "rigor" have been at an all-time high ever since. Statements of all kinds (definitions, axioms, theorems) are scrutinized for the tiniest hidden assumptions, and proofs are made to adhere to the strictest logic. If it all seems a trifle paranoid, think about this: if so simple a concept as "set" has turned out to contain the seed of a paradox, what of most other mathematical concepts, like "number" and "function", that are considerably more complex?

Graphs

Definition 5. A *graph* is an object consisting of two sets called its *vertex set* and its *edge set*. The vertex set is a finite nonempty set. The edge set may be empty, but otherwise its elements are two-element subsets of the vertex set.

Examples. 1) The sets $\{P, Q, R, S, T, U\}$ and $\{\{P,Q\}, \{P,R\}, \{Q,R\}, \{S,U\}\}$ constitute a graph, which we can call "G" for short.

2) $\{1, 2, 3, 4, 5\}$ and $\emptyset$ constitute a second graph, "H".

3) "J" is a graph having vertex set $\{1, 2, 3, 4, 5\}$ and edge set $\{\{1,2\}, \{1,3\}, \{1,4\}, \{1,5\}, \{2,3\}, \{2,4\}, \{2,5\}, \{3,4\}, \{3,5\}, \{4,5\}\}$.

I realize that G, H, and J don't look much like the things in Figure 1, but they will when we start to draw diagrams of graphs.

Definition 6. The elements of the vertex set of a graph are called *vertices* (singular: vertex) and the elements of the edge set are called *edges*. We shall denote the number of vertices by "v" and the number of edges by "e".

Examples. 1) The vertices of the above graph G are P, Q, R, S, T,

'}, $\{P,R\}$, $\{Q,R\}$, and $\{S,U\}$. G has $v = 6$

2) The vertices of H are 1, 2, 3, 4, and 5. It has no edges. For H, $v = 5$ and $e = 0$.

3) J has $v = 5$ and $e = 10$.

Definition 7. If $\{X,Y\}$ is an edge of a graph, we say that $\{X,Y\}$ *joins* or *connects* the vertices X and Y, and that X and Y are *adjacent* to one another. The edge $\{X,Y\}$ is *incident* to each of X and Y, and each of X and Y is *incident* to $\{X,Y\}$. Two edges incident to the same vertex are called *adjacent* edges. A vertex incident to no edges at all is *isolated*.

Examples. In the graph G of previous examples, $\{P,Q\}$ joins P and Q; P and Q are adjacent; P and S are not adjacent; $\{P,Q\}$ is incident to Q; $\{P,Q\}$ is not incident to R; P is incident to $\{P,R\}$; P is not incident to $\{Q,R\}$; $\{P,R\}$ and $\{Q,R\}$ are adjacent; $\{P,R\}$ and $\{S,U\}$ are not adjacent; and T is isolated.

Graph Diagrams

It is customary to represent graphs by drawings, much as it is customary for geometers to represent geometric objects by drawings. For example, Figure 3 is a representation of the graph G we discussed in the last section. The vertices have been represented by heavy dots, and vertex adjacency has been represented by connecting the corresponding dots. Representing graphs by such diagrams makes their structure easier to grasp, but you must be careful not to read too much into the diagram. Diagrams have properties over and above those consequent to their being graph representations. For example, the edges in the picture below each have a certain length, and a specific shape, and they form measurable angles where they meet; and the vertices have specific positions on the page relative to one

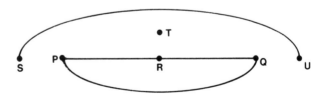

Figure 3

another. But surely these properties were not derived from the abstract definition of *G*, which is merely "*G* is the graph having vertex set {*P, Q, R, S, T, U*} and edge set {{*P,Q*}, {*P,R*}, {*Q,R*}, {*S,U*}}." Rather they are incidental features of the diagram, in fact unavoidable features, but irrelevant insofar as the diagram represents a graph. By many standards Figure 4 is different from Figure 3, but viewed as a graph representation, it is just another picture of *G*. It has the same vertices and the same connections.

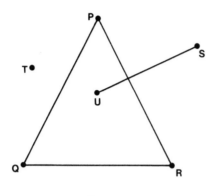

Figure 4

It appears then that graph diagrams can look quite different, yet in some sense be the same.

Definition 8. We will say that two graphs are *equal* if they have equal vertex sets and equal edge sets. And we will say that two graph diagrams are *equal* if they represent equal vertex sets and equal edge sets.

It may seem that this definition is circular, as it defines "equality" in terms of equality. But it is not. Equality of sets is an old concept, introduced in Definition 4, which we are using to define the new concepts of equality of graphs and equality of graph diagrams.

Examples. 1) The graph *G* having vertex set {*P, Q, R, S, T, U*} and edge set {{*P,Q*}, {*P,R*}, {*Q,R*}, {*S,U*}} is equal to the graph *L* having vertex set {*T, P, R, Q, U, S*} and edge set {{*Q,R*}, {*R,P*}, {*U,S*}, {*P,Q*}}; and *G* is equal to any other graph differing from *G* only by a permutation of set elements. I won't belabor the point as you have probably taken it for granted all along. The first part of the definition only makes this understanding explicit.

2) Though the second part of the definition isn't surprising either on an intellectual level, it is interesting visually. By it the two diagrams in Figure 5 are equal. In fact, they and Figures 3 and 4 are all equal as they all represent G; they have vertices P, Q, R, S, T, U and no others, and edges $\{P,Q\}$, $\{P,R\}$, $\{Q,R\}$, $\{S,U\}$ and no others. Apparent differences are irrelevant.

3) The diagrams of Figure 6 are equal. They represent the graph H mentioned in the last section.

4) The diagrams of Figure 7 are equal. They represent J from the last section.

I want to emphasize that to verify that two graph diagrams are equal, it's not necessary to write down their vertex sets and edge sets. All you have to do is check that they have the same number of vertices, that the vertices have the same names, and that the graphs have the same adjacency structure. For example, without writing anything down I can verify that the diagrams of Figure 5 are equal, by saying to myself: "Both graphs have six vertices labeled P, Q, R, S, T, and U. In both graphs P is adjacent to Q and R and to nothing else, Q is adjacent to P and R and to nothing else, R is adjacent to P and Q and to nothing else, S and U are adjacent only to each other, and T is isolated."

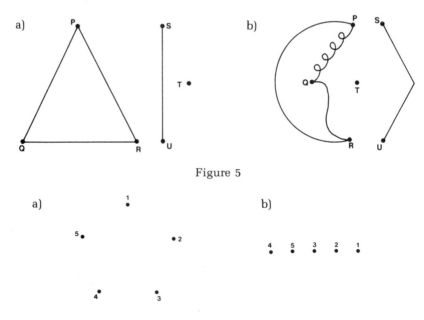

Figure 5

Figure 6

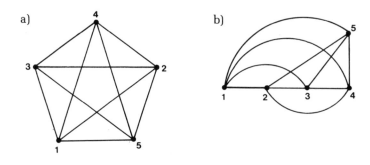

Figure 7

Cautions

1) It so happens that pencil and paper, blackboard and chalk are handy, so graphs are usually represented by flat drawings. But three-dimensional models made of sticks and wads of chewing-gum would do as well. The plane surface is just another incidental feature of a diagram. In a later chapter we will draw graphs on doughnuts!

2) The object in Figure 8 is not a graph. This is because an edge is by definition a set of two vertices, hence cannot exist without a vertex at each end. We have already seen that a graph can have a vertex without incident edges (an isolated vertex); note now that edges, on the other hand, must always have two incident vertices.

3) The graph of Figure 9 is not equal to the graph of Figure 5a, because in Figure 9 there is no edge connecting P to R. Vertex adjacency is not like electric current: it is not transmitted through the intermediary vertex Q.

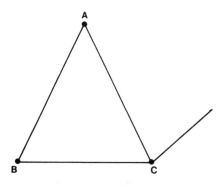

Figure 8

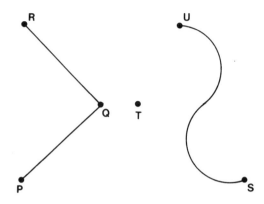

Figure 9

4) The graphs in Figure 7 have a number of edge-crossings. An edge-crossing is not a vertex, but yet another incidental feature of a diagram. Edge-crossings can always be avoided in three-dimensional models, so they are certainly not essential properties of a graph. But even when, as often happens, they cannot be avoided in our standard two-dimensional diagrams, they should not be taken for vertices. To prevent ambiguity we've been using heavy dots for vertices.

5) Sharp corners on edges aren't vertices either; for example in Figure 5b the only vertices are P, Q, R, S, T, and U.

6) The definition of graph precludes "loops" (vertices joined to themselves) and "skeins" (several edges joining the same pair of vertices). This is because a loop would translate abstractly as an "edge" of the form $\{A,A\}$, which is impossible as $\{A,A\}$ is not a set (see p. 14 and Definition 5); and a skein would imply the multiple inclusion of an element $\{A,B\}$ in the edge collection, which would prevent the collection of edges from being a set as required. Were we to allow skeins but not loops we would have something called a "multigraph"; the result of allowing both is called a "pseudograph". We will discuss multigraphs in Chapter 8, but we will not discuss pseudographs. See Figure 10.

7) The edges of a graph are undirected; that is, any impression that the notation $\{A,B\}$ may give of an edge "going from A toward B" is unintentional. Remember that $\{A,B\} = \{B,A\}$. A graph-like thing in which the edges have direction is called a "digraph". We will not discuss digraphs in this book. See Figure 11.

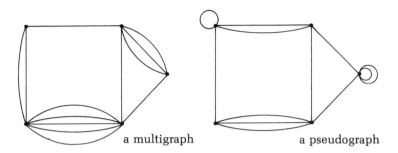

a multigraph   a pseudograph

Figure 10

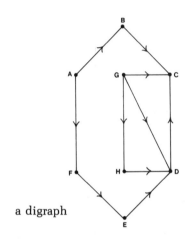

a digraph

Figure 11

Common graphs

Some graphs have acquired more or less standard names because they turn up so frequently.

Definition 9. If v is an integer greater than or equal to 3, the *cyclic graph on v vertices*, denoted "C_v", is the graph having vertex set $\{1, 2, 3, ..., v\}$ and edge set $\{\{1,2\}, \{2,3\}, \{3,4\}, ..., \{(v-1), v\}, \{v,1\}\}$.

Don't let the notation throw you. What we have defined is a family of infinitely many graphs called "cyclic graphs". There is one cyclic graph for each integer, beginning with 3. To find out what a particular member of the family looks like, say C_5, you simply replace all occurrences of "v" by 5 in the definition.

Example. C_5 has vertex set {1, 2, 3, 4, 5} and edge set {{1,2}, {2,3}, {3,4}, {4,5}, {5,1}}. You can draw C_5 yourself.

Three more cyclic graphs have been drawn in Figure 12.

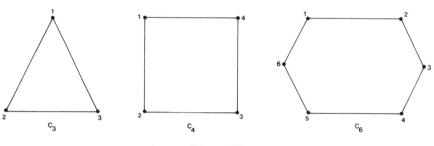

Figure 12

Definition 10. If v is a positive integer, the *null graph on v vertices,* denoted "N_v", is the graph having vertex set {1, 2, 3, ..., v} and no edges.

There are infinitely many null graphs, one for each positive integer. The first three have been drawn in Figure 13. Notice that the graph H of two sections ago has now been given a new name, N_5. All graphs have at least one vertex since Definition 5 requires that a vertex set be nonempty. Thus N_1 is the smallest possible graph, and the only one (essentially) for which $v = 1$.

The next family of graphs is in a sense "opposite" to the null graphs.

Definition 11. If v is a positive integer, the *complete graph on v vertices,* denoted "K_v", is the graph having vertex set {1, 2, 3, ..., v} and all possible edges.

Example. Figure 14 shows three complete graphs. Note that the graph

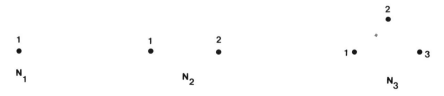

Figure 13

J of two sections ago has just been renamed K_5. K_1 and N_1 are equal, as are K_3 and C_3.

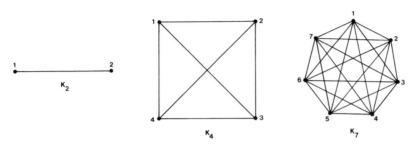

Figure 14

Definition 12. The *utility graph*, denoted "*UG*", is the graph having vertex set $\{A, B, C, X, Y, Z\}$ and edge set $\{\{A,X\}, \{A,Y\}, \{A,Z\}, \{B,X\}, \{B,Y\}, \{B,Z\} \{C,X\}, \{C,Y\}, \{C,Z\}\}$.

Unlike the terms "cyclic graph", "null graph", and "complete graph", that refer to infinite families of graphs, there is only one "utility graph", drawn in Figure 15. The utility graph gets its name from a puzzle in which three houses (A, B, and C) and three utility companies (X, Y, and Z) are represented by dots on a sheet of paper, the task being to connect each house to each utility without crossing any lines. It cannot be done, as we shall see.

Discovery

Figure 16 is a picture of K_{16}. As you can see, the number of edges of a complete graph goes up pretty fast. Given a relatively complicated complete graph, say K_{1000}, it would be helpful if we could determine

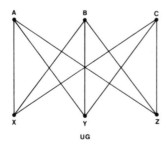

Figure 15

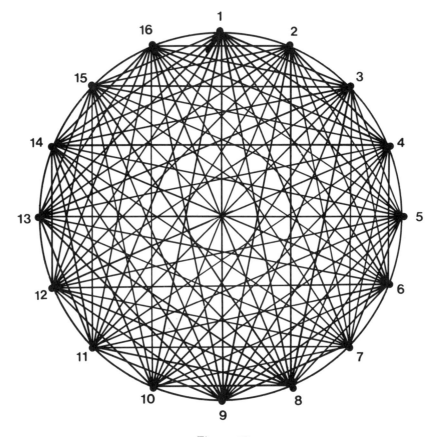

Figure 16

how many edges it has without having to draw it and count them one by one. In this section I will develop a simple formula that does this, and I offer it as an example of how theorems are discovered in the theory of graphs.

On a piece of paper I draw the first few complete graphs. Next, I count their edges and list the totals:

v	e
1	0
2	1
3	3
4	6
5	10

6	15
7	21
8	?

Now I look for a pattern to the numbers in the e column.

I notice one. The e's go up by consecutive integers. From 0 to 1 the jump is 1, from 1 to 3 it is 2, from 3 to 6 it is 3, from 6 to 10 it is 4, etc.

Now I notice that these jumps are the numbers in the first column, so the pattern I've found amounts to this: each e is the sum of the two numbers in the row above it. The 1 in the e column is $1 + 0$, the sum of the row above; similarly the 3 is $2 + 1$, the 6 is $3 + 3$, the 10 is $4 + 6$, etc.

I make a conjecture that my pattern is genuine, that is, that it will continue no matter how far the table is extended. I express my conjecture as an equation.

Conjecture. *For any v,*

(no. edges of K_{v+1}) = (no. vertices of K_v) + (no. edges of K_v).

Of course, the number of vertices of K_v is v, so I can rewrite my conjecture as follows:

Conjecture. *For any v,*

(no. edges K_{v+1}) = v + (no. edges K_v).

To test this I'll try it on K_8, which would have been the next row of the table.

According to my conjecture, (no. edges K_8) = 7 + (no. edges K_7) = 7 + 21 = 28.

Now I draw K_8 and count edges.

K_8 does have 28 edges. I feel pretty confident that my conjecture is right.

But while the conjecture looks good, it's really not what I want. For example, all it tells me about K_{1000} is that (no. edges K_{1000}) = 999 + (no. edges K_{999}), and I don't know the number of edges of K_{999}. Of course (no. edges K_{999}) = 998 + (no. edges K_{998}), but I don't know the number of edges of K_{998} either. I can keep going right down to the beginning, but that doesn't seem any easier than drawing K_{1000} and counting directly. What I need is a pattern that doesn't involve any previous e's.

So I look for such a pattern. After some searching, I find one.

Each e is half the product of its v and the previous v. For example $1 = (1/2)(2 \cdot 1)$, $3 = (1/2)(3 \cdot 2)$, $6 = (1/2)(4 \cdot 3)$, $10 = (1/2)(5 \cdot 4)$, etc. This gives me a new conjecture.

Second conjecture. *For any v,*

$$(\text{no. edges } K_{v+1}) = (1/2)(v + 1)(v).$$

I check if this squares with the first conjecture. The first conjecture said that (no. edges K_{v+1}) = v + (no. edges K_v). By the second conjecture (no. edges K_{v+1}) = $(1/2)(v + 1)(v)$ and (no. edges K_v) = $(1/2)(v)$ $\cdot (v - 1)$; substituting these values into the first conjecture I get

$$(1/2)(v + 1)(v) = v + (1/2)(v)(v - 1)$$
$$(1/2)(v + 1)(v) = (1/2)(2v) + (1/2)(v)(v - 1)$$
$$(1/2)(v + 1)(v) = (1/2)[2v + v(v - 1)]$$
$$(1/2)(v + 1)(v) = (1/2) v[2 + v - 1]$$
$$(1/2)(v + 1)(v) = (1/2) v(v + 1).$$

The last equation is true, so I conclude that my two conjectures are compatible. They are both true or both false. I don't know for a fact that either one is true, but if I'm wrong I have to be doubly wrong.

Furthermore, I notice that the second conjecture, if true, is exactly the sort of thing I want. With it there's no problem to finding the number of edges of K_{1000}: no. edges K_{1000} = $(1/2)(1000)(999)$ = $(1/2) \cdot (999,000)$ = $499,500$. So the second conjecture really seems to be it.

Of course now I have to prove it, if I can.

I try to think of a reason why the second conjecture would have to be true for any complete graph.

I get it by noticing that the number of edges in a graph is half the number of edge-ends. Let me write this up for you formally. My second conjecture is now a theorem.

Theorem 2. *The number of edges in a complete graph K_v is given by the formula $e = (1/2) v(v - 1)$.*

Proof. We will count the number of edge-ends in K_v. K_v has v vertices. Each vertex is joined to the other $v - 1$ vertices, so at each vertex there are $v - 1$ edge-ends. Therefore the total number of edge-ends in K_v is $v(v - 1)$. Every edge has two ends, so the number of edges in K_v is half the number of edge-ends, or $(1/2) v(v - 1)$.

Complements and subgraphs

Heretofore the letters "G", "H", and "J" have been used only in reference to three specific graphs introduced a few sections ago. But from now on we will use them as convenient names for whatever graphs are under discussion, much as an algebraist would use "x", "y", and "z" to denote whatever unknown quantities he or she was considering at the moment.

Definition 13. If G is a graph then the *complement of G,* denoted "$\bar{G}$", is a graph having the same vertex set; the edge set of $\bar{G}$ consists of all two-element subsets of the vertex set which have not already been included in the edge set of G.

Thus $\bar{G}$ has the same vertices as G but the edges are in the "opposite" places. If two vertices are connected in G, they are not connected in $\bar{G}$; and if two vertices are not connected in G, then they are connected in $\bar{G}$.

Examples. 1) The complements of the first three cyclic graphs are shown in Figure 17a, b), and c).

2) Null graphs and complete graphs are complementary. That is, for every positive integer v, $\bar{N}_v$ is equal to K_v and $\bar{K}_v$ is equal to N_v.

3) Figure 17d, e), and f) are all $\overline{UG}$.

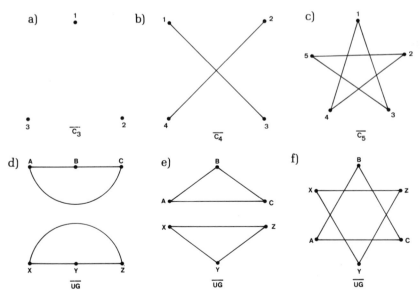

Figure 17

Note that for any graph G, the complement $\bar{G}$ is also a graph, so we may take its complement $\bar{\bar{G}}$; and that when this is done the result is the original graph G.

Definition 14. A graph H is a *subgraph* of a graph G if the vertex set of H is a subset of the vertex set of G and the edge set of H is a subset of the edge set of G.

Examples. 1) In Figure 18 the first five graphs are all subgraphs of the sixth.

2) C_4 is a subgraph of K_5. For that matter, so are $C_3, C_5,$ $N_1, N_2, N_3, N_4, N_5, K_2, K_4,$ and K_5. And there are scores of others that we haven't given names to.

3) C_4 is a subgraph of K_4, which in turn is a subgraph of K_6.

A subgraph is a graph contained in another graph. Note that since every set is a subset of itself, every graph is a subgraph of itself. Intuitively a subgraph is the result of attacking a graph with an eraser, with two qualifications: you needn't erase anything, since every graph is a subgraph of itself; and though you may erase edges as much as you want, you must not erase a vertex without also erasing all edges incident to it, or the result will not be a graph (see item 2 in the "Cautions" section).

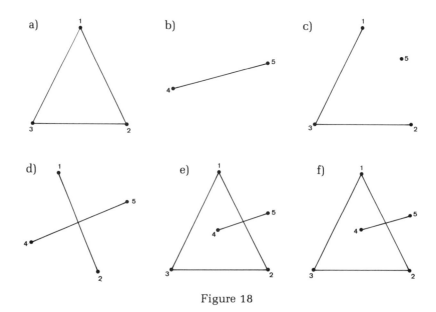

Figure 18

Isomorphism

Something peculiar is going on in Figure 19. The first two graphs are equal, the last two graphs are equal, but the middle two aren't. That hardly seems right, because the only difference between b) and c) lies in how the vertices have been labeled. As we have defined the term "equal", however (see Definition 8), there's no doubt that b) and c) are not equal, for their vertex sets {P, Q, R, S, T, U} and {1, 2, 3, 4, 5, 6} are unequal sets and their edge sets are unequal also.

We obviously need a new concept, one that will make our mathematics sensitive to the kind of "sameness" we intuitively attribute to graphs b) and c). Such a concept exists; it is called "isomorphism", from Greek roots meaning "same structure". The notion of isomorphism is fundamental to the theory of graphs, so I plan to discuss it at some length, beginning with a precise definition. But first a preliminary definition.

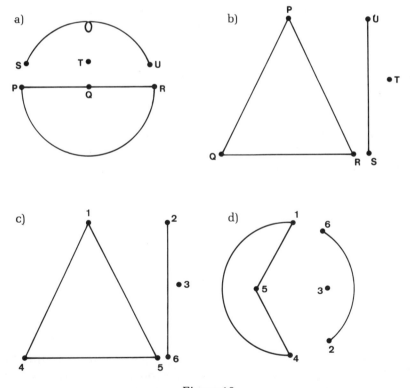

Figure 19

Definition 15. If A and B are sets, then a *one-to-one correspondence* between A and B is an association of elements of A with elements of B in such a way that

    (1) to each element of A there has been associated a single element of B, and

    (2) to each element of B there has been associated a single element of A.

Thus the presence of a one-to-one correspondence between two finite sets means that they at least have the same number of elements. But there is something more. Each element of each set has been made to correspond with a particular element of the other set. So if q is an element of A, one may speak of "the element of B corresponding to q." A simple way of displaying a one-to-one correspondence is to draw arrows between corresponding elements.

Examples. If $A = \{1, 2, 3, 4, 5, 6, 7, 8\}$ and $B = \{\&, +, !, ?, \%, \#, \div, \$\}$, then

(1)

$$1 \leftrightarrow \&$$
$$2 \leftrightarrow +$$
$$3 \leftrightarrow !$$
$$4 \leftrightarrow ?$$
$$5 \leftrightarrow \%$$
$$6 \leftrightarrow \#$$
$$7 \leftrightarrow \div$$
$$8 \leftrightarrow \$$$

is a one-to-one correspondence between A and B, and

(2)

$$1 \leftrightarrow !$$
$$2 \leftrightarrow \#$$
$$3 \leftrightarrow \&$$
$$4 \leftrightarrow ?$$
$$5 \leftrightarrow \div$$
$$6 \leftrightarrow +$$
$$7 \leftrightarrow \$$$
$$8 \leftrightarrow \%$$

is another one.

One-to-one correspondence is the principle behind counting. To count the playing cards in a pack a person removes the first card and says—at least thinks—"one"; then removes the second and says "two"; and so on. The person is associating a number to each card. If the number associated with the last card is, say, "fifty", the person concludes that the pack contains fifty cards. A one-to-one correspondence has been established between the cards and the set {1, 2, 3, ..., 50}.

Definition 16. Two graphs are said to be *isomorphic* if there exists between their vertex sets a one-to-one correspondence having the property that whenever two vertices are adjacent in either graph, the corresponding two vertices are adjacent in the other graph. Such a one-to-one correspondence is called an *isomorphism.* If G and H are isomorphic graphs we denote this by writing "$G \cong H$".

An isomorphism is a special one-to-one correspondence in that it not only associates vertices with vertices but also, in a sense, edges with edges. That is, given an isomorphism between the vertex sets of two graphs under which vertices A and B correspond to vertices X and Y, I am guaranteed that if $\{A,B\}$ is in the edge set of the first graph, then $\{X,Y\}$ is in the edge set of the second; and that if $\{A,B\}$ is not in the edge set of the first graph, then $\{X,Y\}$ is not in the edge set of the second. The isomorphism leaves the edge structure undisturbed: the adjacency or nonadjacency of a pair of vertices implies, respectively, the adjacency or nonadjacency of the corresponding vertices. Mathematicians describe this property by saying that an isomorphism is a one-to-one correspondence between vertex sets that "preserves vertex adjacency."

The definition calls two graphs "isomorphic" if "there exists" an isomorphism; so to *prove* that two graphs are isomorphic, all you have to do is find an isomorphism and present it.

Example. The graphs of Figure 20 are isomorphic, and here is an isomorphism:

$$A \leftrightarrow 3$$
$$B \leftrightarrow 1$$
$$C \leftrightarrow 2$$
$$D \leftrightarrow 4$$

To check that this one-to-one correspondence is an isomorphism, I only have to verify that it preserves vertex adjacency as the definition requires. In the first graph B is adjacent to C; B and C correspond to 1 and 2; and in the second graph 1 is adjacent to 2. In the first graph A is adjacent to C; A and C correspond to 3 and 2; and in the second graph 3 is adjacent to 2. In the first graph A is adjacent to D; A and D correspond to 3 and 4; and in the second graph 3 is adjacent to 4. There are no more adjacencies in either graph, so we have shown that whenever two vertices are adjacent in either one of the graphs, the corresponding two vertices are adjacent in the other graph.

Often there exist several isomorphisms between a pair of isomorphic graphs, though of course the definition requires only that there be one.

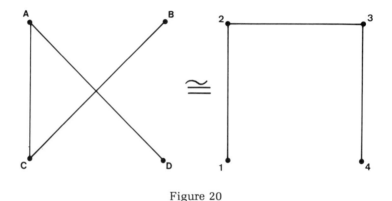

Figure 20

Example. Here is another isomorphism between the graphs of Figure 20:

$$A \leftrightarrow 2$$
$$B \leftrightarrow 4$$
$$C \leftrightarrow 3$$
$$D \leftrightarrow 1$$

You can check for yourself that this one-to-one correspondence preserves vertex adjacency and so is an isomorphism as claimed.

Here's another way of displaying an isomorphism, one that I find a lot simpler. Pick one of the two graphs and put a pair of parentheses next to each vertex. Then fill each pair of parentheses with the name of the corresponding vertex from the other graph.

Examples. 1) The graphs of Figure 21 are isomorphic, and I have indicated an isomorphism by adding parenthetical labels to the second graph; that is, 3 corresponds to D, 2 to C, 4 to A, and 1 to B. To

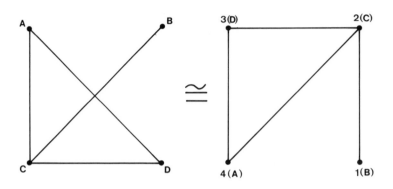

Figure 21

verify that this one-to-one correspondence between the vertex sets is an isomorphism, all I have to do is check that the adjacency of the parenthetical labels in the second graph is the same as their adjacency in the first graph. For instance in the second graph, A is adjacent only to D and C; and in the first graph, A is adjacent only to D and C. So A checks; A's structural involvement is the same in both graphs.

Now I'll do the same for B, C, and D. In the second graph, B is adjacent only to C; and in the first graph, B is adjacent only to C. Check.

In the second graph, C is adjacent to all the other vertices; and in the first graph, C is adjacent to all the other vertices. Check.

In the second graph, D is adjacent only to A and C; and in the first graph, D is adjacent only to A and C. Check.

Everything checks, so the correspondence indicated by the labels in parentheses is an isomorphism, and the graphs of Figure 21 are isomorphic.

2) The graphs of Figure 22 are isomorphic, and I have displayed an isomorphism by the labels added parenthetically to the second graph. It is an isomorphism because, in both graphs:

1 is adjacent only to 3 and 6,
2 is adjacent only to 4 and 7,
3 is adjacent only to 1 and 5,
4 is adjacent only to 2, 5, and 6,
5 is adjacent only to 3, 4, 6, and 7,
6 is adjacent only to 1, 4, and 5, and
7 is adjacent only to 2 and 5.

Therefore, whenever two vertices are adjacent in either graph, the corresponding two vertices are adjacent in the other graph, and we have an isomorphism.

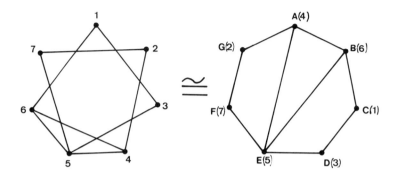

Figure 22

Looking at the second graph in Figure 22, we can think of the labels in parentheses as being new labels that we are about to substitute in order to transform the second graph into a graph equal to the first graph. This suggests an alternate way of defining isomorphism: two graphs are isomorphic if either

    1) they are equal, or

    2) they could be made equal by changing the way one of them is labeled.

By the way, graphs don't *become* isomorphic when you discover an isomorphism. They are of themselves isomorphic, or not, whether or not anyone discovers the fact, or proves it.

Example. The graphs of Figure 23 are isomorphic as they stand, even though under the actual labeling they are not equal (because the edge sets are not equal), and I offer no alternate labeling under which they would be equal. They are isomorphic because they have the *potential* to be labeled so that edges appear in corresponding places.

This apparent independence of mathematical facts from what people do, this sense that mathematics is discovered rather than created, has caused a number of mathematicians (for instance Hardy in *A Mathematician's Apology*) to ascribe to mathematical objects a separate existence on another level of reality, akin to Plato's "world of Ideas". Someone taking this view to its extreme would maintain that the graphs of Figure 23 have been isomorphic since before the formation of the solar system.

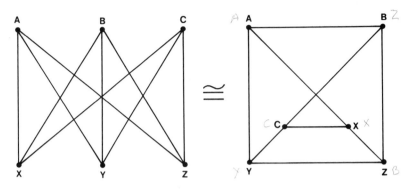

Figure 23

Recognizing isomorphic graphs

   In doing graph theory, it's important not only to understand what it means when someone tells you two graphs are isomorphic, but also to develop some degree of skill at recognizing isomorphic graphs on your own. Often you'll need to decide whether or not two unfamiliar graphs are isomorphic, and then be able to prove that your choice is correct. Unfortunately, recognizing isomorphic graphs, finding specific isomorphisms, and proving graphs to be nonisomorphic are skills that can't really be taught; they come mostly with practice. Nevertheless in this section I will present some general guidelines that I hope will be of help.

Steel balls and rubber bands. This first suggestion is intended primarily for people with a good intuitive grasp of spatial relationships. Think of a graph as a network of steel balls (vertices) and rubber bands (edges). We assume the balls will remain in whatever position we place them and that the rubber bands will never break. Under this interpretation, isomorphic graphs are graphs that can be rearranged to look like one another.

Example. The graphs of Figure 24 are isomorphic. The first can be rearranged to look like the second, as shown in Figures 25 and 26. Start (Figure 25) by pulling triangle 1-5-3 down and away from triangle 6-2-4. This inverts edge {1,4} and stretches edges {5,2} and {6,3}. Now with edge {1,4} as an axis (Figure 26), flip triangle 1-5-3 over so that vertices 5 and 3 exchange places. This twists edge {1,4} and untangles edges {5,2} and {6,3}. The result looks like the second

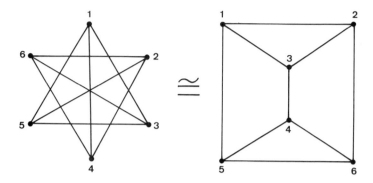

Figure 24

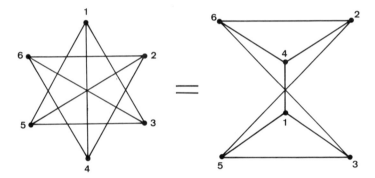

Figure 25

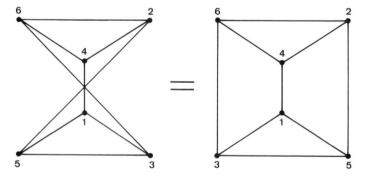

Figure 26

graph in Figure 24, so we can conclude that the graphs in Figure 24 are isomorphic. Moreover, we have discovered a specific isomorphism between the original graphs, which I have indicated by parenthetical labels in Figure 27. I obtained it by associating the labels of the second graph in Figure 26 to those of the second graph in Figure 24.

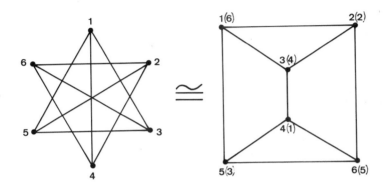

Figure 27

Properties preserved by isomorphism. This is for everyone. We have seen that isomorphism preserves vertex adjacency; it preserves a number of simpler properties as well, of which we will consider four.

i) *The number of vertices.* An isomorphism is a one-to-one correspondence between vertex sets and vertex sets are finite (see Definition 5), so isomorphic graphs must have the same number of vertices.

ii) *The number of edges.* An isomorphism induces a one-to-one correspondence between edge sets and edge sets are finite too (because the vertex sets are finite), so isomorphic graphs must have the same number of edges.

Before considering the next two properties preserved under isomorphism, we need a definition.

Definition 17. The *degree* of a vertex is the number of edges incident to it.

Examples. 1) The graph of Figure 28 has two vertices (H and I) of degree 0, two vertices (E and G) of degree 1, three vertices (A, D, and F) of degree 2, and two vertices (B and C) of degree 3.

2) Each vertex of K_{153} has degree 152.

iii) *The distribution of degrees.* In isomorphic graphs corresponding vertices have the same degree, because isomorphism preserves vertex

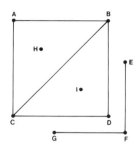

Figure 28

adjacency. Consequently if a graph has, say, one vertex of degree 1, five vertices of degree 3, and two vertices of degree 4, then any graph isomorphic to it will have one vertex of degree 1, five vertices of degree 3, and two vertices of degree 4.

iv) *The number of "pieces" of a graph.* I want to defer a precise definition until later (p. 113), but I think you'll get my meaning if I tell you that the two graphs of Figure 27 are each "in one piece" and the graph of Figure 28 is "in four pieces". At first glance the graph of Figure 17f might appear to be "in one piece", but it's really "in two pieces" as it is equal to the graph of Figure 17e. Since isomorphism preserves vertex adjacency it preserves the number of pieces of a graph, that is, isomorphic graphs must be composed of the same number of pieces.

These properties are useful in two ways. First, they can be used to prove that graphs are *not* isomorphic. Since all four properties are preserved by isomorphisms, a pair of graphs differing in any one of those four respects cannot possibly be isomorphic.

Examples. 1) If the graphs of Figure 29 were isomorphic, they would have the same number of vertices. The first has $v = 6$ and the second has $v = 5$, so they are not isomorphic. One reason is enough, but here are two more: their degree distributions are different, for example the first has two vertices of degree 1 and the second has only one vertex of degree 1; and they have different numbers of pieces, the first being in two pieces while the second is in one piece.

2) The graphs of Figure 30 are not isomorphic because they have different numbers of edges: the first has $e = 6$ and the second has $e = 5$. Also, the first has a vertex of degree 4 and the second has no vertex of degree 4. They have the same number of vertices, and are both in one piece.

3) The graphs of Figure 31 are not isomorphic because they have

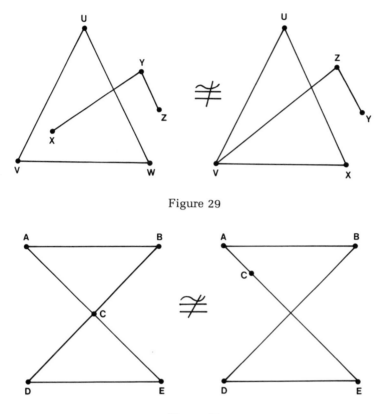

Figure 29

Figure 30

different degree distributions: the first has one vertex of degree 4, two of degree 5, and four of degree 6, whereas the second has no vertices of degree 4, four of degree 5, and three of degree 6. They share the other three properties: they have the same v's and e's and they are both in one piece.

4) The graphs of Figure 32 have the same number of vertices (6), the same number of edges (6), and the same distribution of degrees (all vertices have degree 2). Yet they are not isomorphic because the first is in one piece and the second is in two pieces.

The second use to which the four properties can be put is that they can provide the basis for a conjecture that two graphs *are* isomorphic. Even if a pair of graphs share all four properties, they are not *necessarily* isomorphic; but having checked that they do not differ in these four obvious ways, it is at least less of a gamble to invest some time searching for an isomorphism.

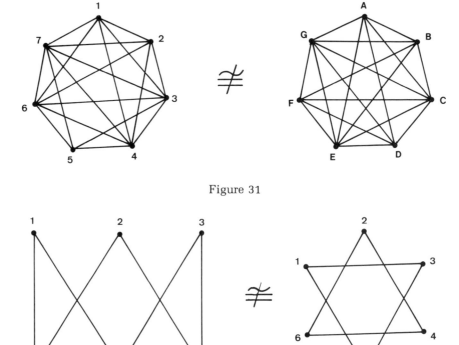

Figure 31

Figure 32

Example. The two graphs in Figure 33 each have $v = 8$ and $e =$
12; they have the same distribution of degrees, namely two vertices
of degree 1, three of degree 3, two of degree 4, and one of degree
5; and they are both in two pieces. This covers the more obvious
possible differences, so I'll conjecture that they are isomorphic and
try to construct an isomorphism. (If I fail I'll come back and search
for differences that are less obvious.) Clearly vertices 4 and 5 of the
first graph would have to correspond to vertices 2 and 3 of the second;
I've indicated this by adding parenthetical labels to the second graph
in Figure 33. And the vertex of degree 5 in the first graph would
have to correspond to the vertex of degree 5 in the second, so after
the label "4" in the second graph I'll put "(3)". The two vertices
of degree 4 in the first graph have to correspond to the two vertices
of degree 4 in the second, but in what order? Should I associate

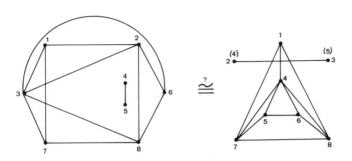

Figure 33

2 with 7 and 8 with 8, or should it be 2 with 8 and 8 with 7? It might make a difference. I'll try 2 with 7 and 8 with 8; if it doesn't work out I'll return to this point and try it the other way. So after labels "7" and "8" in the second graph I'll put "(2)" and "(8)". Now all I have left is to associate the vertices of degree 3. I'll start with vertex 1 in the first graph. I won't match it with the "1" in the second graph, because the two "1"s are involved differently in the structures of their graphs: in the first graph vertex 1 is adjacent to vertices of degrees 3, 4, and 5, but in the second graph vertex 1 is adjacent to vertices of degrees 4, 4, and 5. Instead I'll match the "1" of the first graph with vertex 5 of the second graph, because in the second graph vertex 5 is adjacent to vertices of degrees 3, 4, and 5, a structural involvement like that of vertex 1 in the first graph. So next to the label "5" in the second graph I'll write "(1)". Finally I'll associate vertex 6 of the first graph with vertex 1 of the second, as they are both adjacent to vertices of degrees 4, 4, and 5; this leaves vertex 7 of the first graph associated with vertex 6 of the second graph, as they are the only vertices left. I have displayed all this in Figure 34, and you can verify for yourself that the correspondence does turn out to be an isomorphism. All you have to do is check that the numbers in parentheses in the second graph are connected to one another in exactly the same way as the numbers in the first graph.

The sort of thing we've just done is very often possible. Two graphs sharing all four properties are frequently isomorphic, and if they are an isomorphism can always be found by a careful search. Unfortunately, things don't always turn out so nicely. Graphs sharing the four properties can still be nonisomorphic, making the most patient search for an isomorphism a complete waste of time. This uncertainty puts a little zip into the graph game.

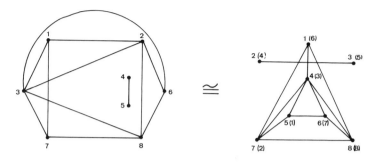

Figure 34

Example. The graphs of Figure 35 are not isomorphic, even though they share the four properties: in each $v = 8$ and $e = 10$; each has four vertices of degree 2 and four of degree 3; and they are both in one piece. The structural difference is that the vertices of degree 2 are not related in the same way in the two graphs. In the first graph the vertices of degree 2 come in adjacent pairs: B is adjacent to H and D is adjacent to F. But in the second graph they are completely separated from one another by vertices of degree 3: no vertex of degree 2 is adjacent to any other vertex of degree 2. To see how this prevents the graphs from being isomorphic, we shall systematically try to construct an isomorphism in all possible ways. We begin by noticing that B, being of degree 2, would have to correspond to 1, 3, 6, or 8.

First case: B corresponds to 1.

Then H, being adjacent to B, would have to correspond to 2 or

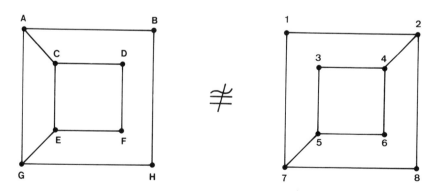

Figure 35

7. But *H* must correspond to a vertex having the same degree, so *H* cannot correspond to either 2 or 7. Therefore no isomorphism is possible if *B* is made to correspond with 1.

Second case: B corresponds to 3.

Then *H*, being adjacent to *B*, would have to correspond to either 4 or 5. But *H* must correspond to a vertex having the same degree, so *H* cannot correspond to either 4 or 5. Therefore no isomorphism is possible if *B* is made to correspond to 3.

Third Case: B corresponds to 6.

Then *H*, being adjacent to *B*, would have to correspond to either 4 or 5, which is impossible as we noted in the second case. Therefore no isomorphism is possible if *B* is made to correspond to 6.

Fourth case: B corresponds to 8.

Then *H*, being adjacent to *B*, would have to correspond to either 2 or 7, which is impossible as we noted in the first case. Therefore no isomorphism is possible if *B* is made to correspond to 8.

We have seen that no isomorphism is possible if *B* is made to correspond to any of 1, 3, 6, or 8. But under any isomorphism, *B* would have to correspond to 1, 3, 6, or 8. Therefore no isomorphism is possible, and the graphs of Figure 35 are not isomorphic.

Semantics

In graph theory there are two concepts of "sameness" whereby graphs are judged to be "the same", namely equality and isomorphism. Of these isomorphism is the more fundamental, and the one with which we will be chiefly concerned from now on. It's true that we may care to know, occasionally, whether or not two graphs are equal, but I introduced the notion of equality mostly as a stepping-stone to the more abstract notion of isomorphism.

An indication of the pervasive role isomorphism has in graph theory is the fact that isomorphism has virtually captured the word "is". In most contexts graph theorists use "is" not to mean "is equal to", as you would naturally think, but to mean "is isomorphic to". This is a departure from the usage of school mathematics, and you may find it takes some getting used to. For example a graph theorist would say "the graph of Figure 36 is the utility graph", by which he or she would mean that it is isomorphic to the utility graph, though by standard usage the graph of Figure 36 most certainly "is not" the utility graph (it is not equal to the utility graph) because its vertex set {1, 2, 3, 4, 5, 6} is different from the vertex set of the utility graph, which (see Definition 12) is {A, B, C, X, Y, Z}.

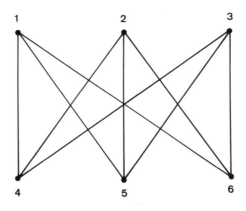

Figure 36

Henceforth we will adopt this semantic convention and use "is" to mean "is isomorphic to". It may seem that by so doing we are blurring an important distinction, but that's really not the case. In the rest of this book we will discuss only those properties of a graph that are shared by all the graphs isomorphic to it, whether or not they are also equal to it. We still have the distinction between equality and isomorphism to fall back on should it ever be important to maintain, but in the meantime letting "is" mean "is isomorphic to" will emphasize isomorphism as the primary sense in which we will consider graphs to be "the same".

Accordingly, from now on whenever we say "this graph is that graph" without qualification, it will mean "this graph is isomorphic to that graph; we don't care if they are equal as well".

Examples. 1) I say, "the first graph in Figure 37 is the complement of UG." It is not *equal* to the complement of UG (the second graph in Figure 37), because its vertex set {G, H, I, J, K, L} is different from the vertex set of the complement of UG, which (by Definitions 12 and 13) is {A, B, C, X, Y, Z}. However, the two graphs in Figure 37 are *isomorphic*, which is what I meant.

2) I say, "UG is a subgraph of K_6." It is not *equal* to a subgraph of K_6, because its vertex set {A, B, C, X, Y, Z} is not a subset of

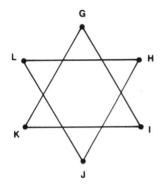

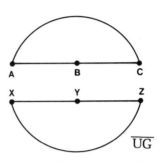

Figure 37

the vertex set of K_6, which is $\{1, 2, 3, 4, 5, 6\}$. (See Definitions 11 and 14.) However, *UG* is *isomorphic* to a subgraph of K_6, for instance the one drawn in Figure 36, and this is what I meant. You can check for yourself that the graph of Figure 36 is equal to a subgraph of K_6 by drawing K_6 and then erasing edges $\{1,2\}$, $\{1,3\}$, $\{2,3\}$, $\{4,5\}$, $\{4,6\}$, and $\{5,6\}$.

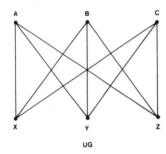

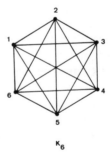

Figure 38

One final thing. Since isomorphism emancipates us from specific labelings, we may as well, whenever convenient, do away with labels altogether. Consequently we will often represent graphs by unlabeled diagrams.

Example. We will speak of the graphs in Figure 39 as respectively K_2, C_5, and N_3 because they are respectively isomorphic to K_2, C_5, and N_3 as originally defined.

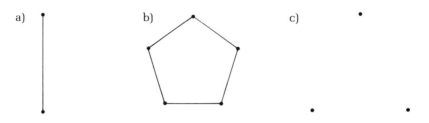

Figure 39

The number of graphs having a given *v*

I want to tell you something of what's known about the size of the "ballpark" in which we are playing our game.

If equality were the criterion for judging graphs to be "the same", and therefore the criterion for judging them to be "different", we would have to say that there "are" an infinite number of graphs with, say, $v = 2$ and $e = 1$, as such a graph can be labeled in infinitely many ways (Figure 40). To a mathematician, who by training and inclination probes for basic structure, this would be a ridiculous situation. Labels are ephemeral; they tell you about as much about the structure of a graph as the list of colors tells you about the structure of a painting.

Isomorphism is the criterion instead, because it is in accord with the mathematician's perception that, for graphs with two vertices and one edge, there is only one basic design. Every graph with $v = 2$ and $e = 1$ is isomorphic to K_2 (the first graph in Figure 39); we will express this by saying "up to isomorphism, K_2 is the only graph with $v = 2$ and $e = 1$", or more simply "K_2 is the only graph with $v = 2$ and $e = 1$", it being understood that we judge graphs to be "different" only if they are not isomorphic.

Similarly, as every graph with two vertices and no edges is isomorphic

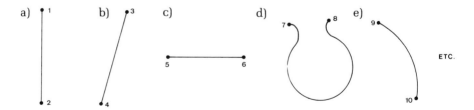

Figure 40

to N_2, N_2 is the only graph with $v = 2$ and $e = 0$. So in all there are two graphs having $v = 2$.

Of course all graphs with $v = 1$ are isomorphic to K_1 (also named N_1), so there is only one graph with $v = 1$.

You can check for yourself that no two of the graphs in Figure 41 are isomorphic, and that every graph with three vertices is isomorphic to one of them. Therefore those four are the only graphs having $v = 3$.

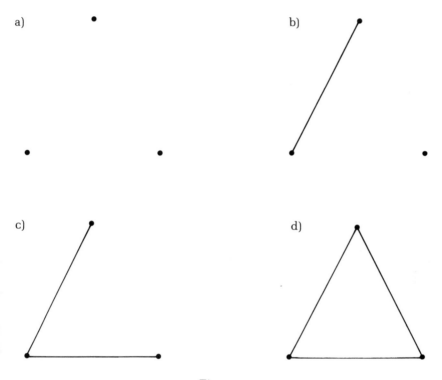

Figure 41

Let's determine how many graphs there are with $v = 4$. The situation is more complicated than $v = 1$, $v = 2$, or $v = 3$, so I'll proceed systematically.

On a piece of paper I arrange seven columns headed by "$e = 0$", "$e = 1$", "$e = 2$", and so on, up to "$e = 6$". A graph with four vertices can have no more than six edges, as complete graphs have all possible edges and K_4 has $e = 6$ by Theorem 2.

In the "$e = 0$" column I draw N_4. Obviously any other graph with

$v = 4$ and $e = 0$ will be isomorphic to N_4, so that's all for that column.

Now I jump to the "$e = 6$" column and draw K_4, which is the complement of N_4. I do the chart by complements because it has occurred to me (think about this) that isomorphic graphs must have isomorphic complements, and nonisomorphic graphs must have nonisomorphic complements.

Back to the "$e = 1$" column. It's clear to me that this column will contain only one graph. (If you're not convinced, draw lots of graphs with $v = 4$ and $e = 1$ and then prove they're all isomorphic.)

Now I'll put the complement of the "$e = 1$" graph under "$e = 5$". I'm confident there is no other graph with $e = 5$, reasoning as follows: if there were two nonisomorphic graphs with $e = 5$, then their complements would be nonisomorphic graphs with $e = 1$; but I'm satisfied there's only one graph with $e = 1$.

A graph with $v = 4$ and $e = 2$ can be constructed in two ways, with the edges separate or joined. So under "$e = 2$" I draw the two graphs of Figure 42. They are not isomorphic, and every graph with $v = 4$ and $e = 2$ is isomorphic to one of them.

Now in the column headed "$e = 4$" I draw the complements of the graphs in Figure 42.

The "$e = 3$" column is the only one left. This column isn't paired with a complementary column because there are an odd number of columns; if a graph has $v = 4$ and $e = 3$, then its complement also has $v = 4$ and $e = 3$. I start with the first graph of Figure 42, and add an edge in all possible ways (Figure 43). These four graphs are all isomorphic, so they amount to only one graph for the "$e = 3$" column. Next I take the second graph of Figure 42, and add an edge in all possible ways (Figure 44). Of these four graphs the first and

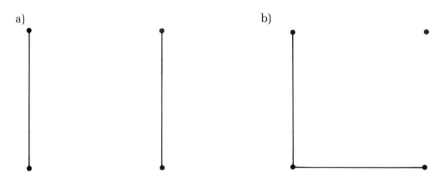

a)   b)

Figure 42

(a)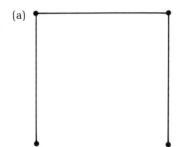

b)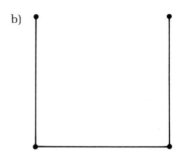

c)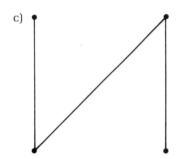

(d)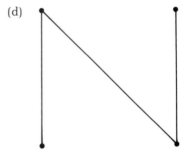

Figure 43

last are nothing new, as they are isomorphic to one another and to
the graphs of Figure 43. But the middle two are new: they are not
isomorphic to each other or to the graph of Figure 43. I conclude
that there are three patterns for a graph with $v = 4$ and $e = 3$.

My chart is finished. I have reproduced it in Figure 45. There are
eleven graphs with $v = 4$.

If you're wondering where the "square" is, it's the first graph in
the "$e = 4$" column. It's twisted because of the particular drawing
I made of its complement, which is the first graph under "$e = 2$".
If not having a recognizable square bothers you, you can redo the
first graph under "$e = 2$" to look like an X.

By the way, some people get the impression that the graphs we've
given specific names to are the only graphs there are. If you are one
of these people, please notice that of the graphs in Figure 45 only
three have been named—N_4, C_4, and K_4.

a)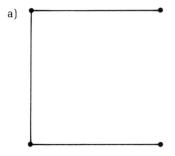

b)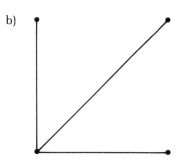

c)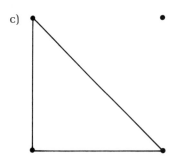

d)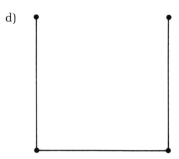

Figure 44

You may wonder how many graphs there are with $v = 5$, $v = 6$, etc. The totals get large very fast:

v	number of graphs
1	1
2	2
3	4
4	11
5	34
6	156
7	1,044
8	12,346
9	308,708

These numbers are computed using something called the "Polya Enumeration Theorem", which is too difficult to take up in this book.

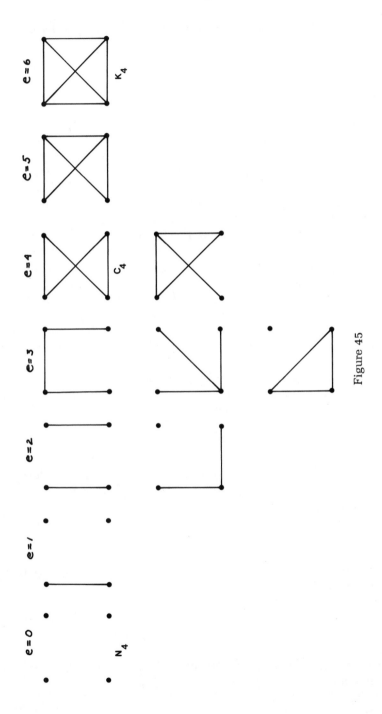

Figure 45

There is no simple way of determining the number of graphs having a given v.

Exercises

1. List all subsets of the set $\{1, 2, 3\}$. There are eight of them.
2. It follows from the Law of Excluded Middle (page 18) that to prove a mathematical statement true, it suffices to show that it cannot be false. Letting A be a set and J an empty set, show that the statement "J is a subset of A" cannot be false and so is true. This proves that an empty set is a subset of every set and promotes the Convention on page 14 to Theorem.
3. This is another version of Russell's Paradox. The village barber shaves those and only those men who live in the village and do not shave themselves. The village barber is a man and he lives in the village.

   Consider the question "Who shaves the barber?" Then explain how this situation is equivalent to Russell's Paradox.
4. Let "S" be the collection of all sets that can be described in an English sentence of twenty-five words or less. Is S a set? Why or why not?
5. If v is an integer greater than or equal to 2, the *path graph on v vertices*, denoted "P_v", is the graph having vertex set $\{1, 2, 3, . . ., v\}$ and edge set

$$\{\{1,2\}, \{2,3\}, \{3,4\}, ...,\{v-1,v\}\}.$$

   Draw the first five path graphs. Then find and prove a formula analogous to Theorem 2 for the number of edges of P_v.
6. If v is an integer greater than or equal to 4, the *wheel graph on v vertices*, denoted "W_v", is the graph having vertex set $\{1, 2, 3, ..., v\}$ and edge set $\{\{1,2\}, \{1,3\}, ..., \{1,v\}, \{2,3\}, \{3,4\},$ ..., $\{v-1,v\}, \{v,2\}\}$. Draw the first five wheel graphs. Then find and prove a formula analogous to Theorem 2 for the number of edges of W_v.
7. Use Theorem 2 to prove that

$$1 + 2 + ... + (v-1) = (1/2)v(v-1)$$

   for any integer v greater than or equal to 2. Do not use any arithmetic or algebra.

8. Let G be a graph with v vertices and e edges. In terms of v and e, how many edges has $\bar{G}$?

9. Prove: if a graph G has $v = 6$ then G or $\bar{G}$ (possibly both) has a subgraph isomorphic to K_3.

10. Use Exercise 9 to prove that in any gathering of six people there are either three people who are mutually acquainted or three people who are mutually unacquainted, possibly both.

11. Prove: the sum of the degrees of the vertices of a graph is $2e$.

12. Use Exercise 11 to answer these questions:
    a) If a graph has $v = 9$, 4 vertices of degree 3, 2 vertices of degree 5, 2 of degree 6 and 1 of degree 8, how many edges has the graph?
    b) UG has $v = 6$ and every vertex has degree 3; prove that there are however no graphs with $v = 7$ in which every vertex has degree 3.

13. Prove that $C_5 \cong \bar{C}_5$. Then prove that no other cyclic graph is isomorphic to its complement.

14. Prove: If $G \cong \bar{G}$ then v or $v - 1$ is a multiple of 4. (You might use Theorem 2 and the fact that isomorphic graphs have the same number of edges.)

15. If $G \cong \bar{G}$, G is called a "self-complementary" graph. Exercise 13 says that C_5 is self-complementary and is the only cyclic graph that is. Find two other self-complementary graphs. (You might use Exercise 14 to narrow your search.)

16. Find a self-complementary graph with $v = 8$. Of the 12,346 graphs with $v = 8$ only four are self-complementary.

17. Since there are an odd number of graphs having $v = 4$ (drawn in Figure 45), one of them must be self-complementary. Which one?

18. How many different one-to-one correspondences are there between $\{a, b, c, d, e, f, g, h, i, j\}$ and $\{1, 2, 3, 4, 5, 6, 7, 8, 9, 10\}$? (Two one-to-one correspondences are "different" if there is at least one element that they associate with different elements.)

19. We said in the text that the existence of a one-to-one correspondence between two finite sets implies that they contain the same number of elements. The situation for infinite sets is somewhat different. The set of positive integers $\{1, 2, 3, ...\}$ in some sense contains "twice as many" elements as the set of even positive integers $\{2, 4, 6, ...\}$, yet it is possible to put the two sets into a one-to-one correspondence. Do so.

20. Besides the four properties mentioned in the text, another property preserved by isomorphism is the distribution of subgraphs. That

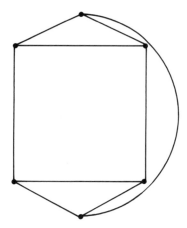 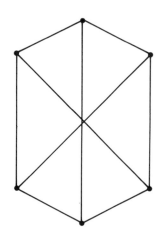

Figure 46

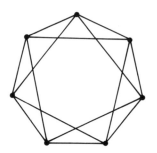 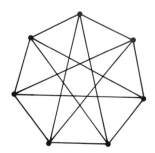

Figure 47

a) 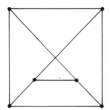					b) 					c)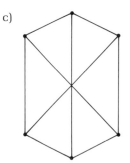

Figure 48

is, if two graphs are isomorphic and you select a subgraph at random from either one, then the other will necessarily have an isomorphic subgraph. Hence you can prove that two graphs are not isomorphic by finding a subgraph of one that is not a subgraph of the other. Use this fact to devise a proof, shorter than the one in the text, that the graphs of Figure 35 are not isomorphic.

21. K_3 (with its vertices labeled) has 17 unequal subgraphs. Draw them.

22. The number of *nonisomorphic* subgraphs of K_3 is only 7. Draw them.

23. The graphs of Figure 46 are not isomorphic. Prove this by finding a subgraph of one that is not a subgraph of the other.

24. Satisfy yourself that isomorphic graphs have isomorphic complements and that consequently nonisomorphic graphs have nonisomorphic complements. Then use this fact to devise a proof, different from the one in the text, that the graphs of Figure 31 are not isomorphic.

25. Use the technique of Exercise 24 to prove that the graphs of Figure 46 are not isomorphic.

26. Draw all graphs having $v = 5$. There are 34 of them. (Imitate the procedure I used in the text to find all graphs with $v = 4$.)

27. In the table on page 54, the numbers in the second column are mostly even. If we ignore the first row on the ground that $v = 1$ is such a trivial situation that its uniqueness is unremarkable, that leaves $v = 4$ as the only number of vertices listed for which there are an odd number of graphs. Do you think this is due to chance, or can you think of some reason why $v = 4$ should be unique? If the table were continued, do you think more odd numbers would turn up in the second column?

28. Prove that the graphs of Figure 47 are isomorphic.

29. Prove that the graphs of Figure 48 are all isomorphic to the utility graph.

30–37. In each of Figures 49–56, decide whether or not the two graphs are isomorphic. If you decide they are isomorphic, prove it by finding a labeling under which they would be equal. If you decide they are not isomorphic, prove it by finding a property which is preserved by isomorphism but in terms of which the two graphs differ. (We now have six such properties: the four mentioned in the text, and two more introduced in Exercises 20 and 24.)

38. A chess tournament consists of twenty-five players, each of whom

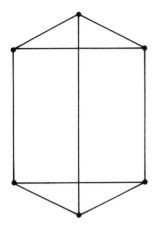

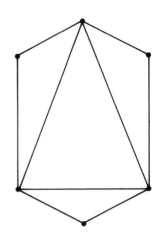

Figure 49

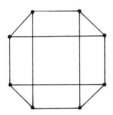
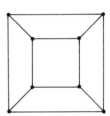

Figure 50

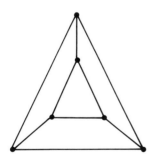
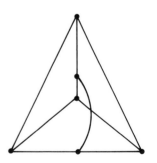

Figure 51

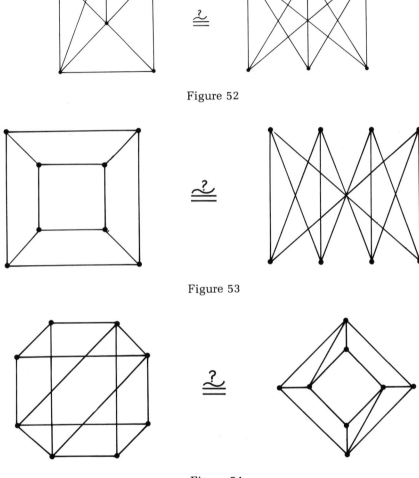

Figure 52

Figure 53

Figure 54

plays one game with every other player. How many games are played during the tournament?

39. When I said on page 56 "There is no simple way of determining the number of graphs having a given v," I meant of course that there is no simple way of determining the number of *nonisomorphic* graphs having a given v. Prove that there are exactly $2^{(1/2)v(v-1)}$

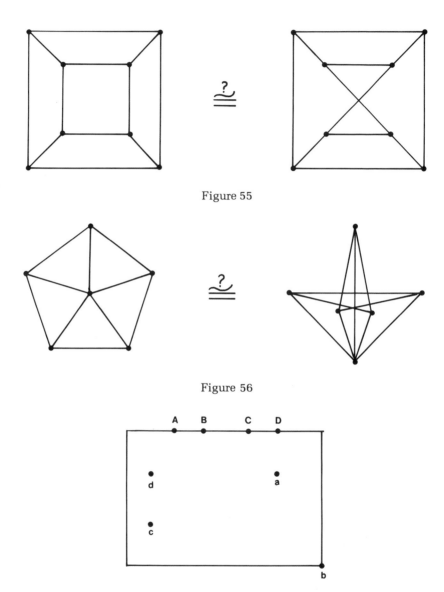

Figure 55

Figure 56

Figure 56A

unequal graphs whose v vertices have been labeled with the integers from 1 to v.

40. The utility puzzle mentioned on page 27 was invented by Henry Ernest Dudeney (1857–1930), a self-taught mathematician who was

the foremost puzzlemaker of his time. Here is another of Dudeney's puzzles, one which, unlike the utility puzzle, has a solution. In figure 56A connect *a* to *A*, *b* to *B*, *c* to *C*, and *d* to *D* by edges which are inside the rectangle and don't cross one another.

Suggested reading

On Sets

* *One, Two, Three … Infinity: Facts and Speculations of Science* by George Gamow (Viking, 1961), chapter 1.

* *Stories About Sets* by N. Ya. Vilenkin (Academic Press, 1968), chapters 1–3.

On Paradoxes

* "Paradox Lost and Paradox Regained" by Edward Kasner and James R. Newman, reprinted in volume 3 of *The World of Mathematics*, edited by James R. Newman (Simon and Schuster, 1956).

* "Paradox" by W. V. Quine in the April, 1962 issue of *Scientific American*; reprinted as chapter 28 of *Mathematics in the Modern World: Readings from Scientific American* with introductions by Morris Kline (W. H. Freeman, 1968).

On Bertrand Russell

"Bertrand Russell 1872–1970" in the February 16, 1970 issue of *Newsweek*. A zippy obituary.

On Logic

* "What the Tortoise Said to Achilles" by Lewis Carroll, reprinted in volume 4 of *The World of Mathematics*, edited by James R. Newman (Simon and Schuster, 1956).

On Mathematical Discovery

* *How to Solve It: A New Aspect of Mathematical Method* by G. Polya (Princeton University Press, 1957). Handy when you're stuck.

3. PLANAR GRAPHS

Introduction

Definition 18. A graph is *planar* if it is isomorphic to a graph that has been drawn in a plane without edge-crossings. Otherwise a graph is *nonplanar*.

(I have dropped the distinction between an abstractly given graph and a graph diagram. I use the word "graph" to refer to both. When I say "a graph that has been drawn, etc.," I obviously mean a graph diagram.)

Thus a planar graph, when drawn on a flat surface, either has no edge-crossings or can be redrawn without them. The three graphs of Figure 57 are planar, the first two because they have no edge-crossings and the third because its edge-crossing is avoidable—the third is isomorphic to the second.

In principle it is a simple matter to demonstrate planarity. Like isomorphism, planarity is a potential. A graph is planar if a certain task, the drawing of the graph without edge-crossings, is possible; and we can show the task to be possible simply by performing it. In practice this is not always easy. For example, although the graph of Figure 58a is planar, you might require several tries in order to draw it without edge-crossings.

But in principle, as well as practice, it is difficult to demonstrate nonplanarity. A graph is nonplanar if drawing the graph without edge-crossings is impossible, so we are required to prove the impossibility of performing that task—a tall order.

Example. The graphs of Figure 58b and c) are nonplanar, but for now you will have to take my word for it. You might try to draw these graphs without edge-crossings and if you do your failures will incline you to believe me. But you should not allow yourself to be convinced by repeated failures, for how can you be sure you just haven't been clever enough to make a crossing-free drawing? Of course,

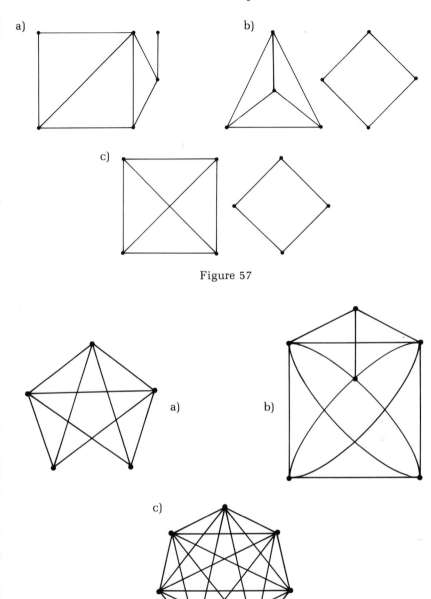

Figure 57

Figure 58

you can systematize your attempts and draw edges in all possible ways, but for any moderately complicated graph the number of possibilities is quite large, in which case you might overlook one. In general the claim that a graph is nonplanar needs backing up by rigorous proof.

Assuming for the moment that the graphs of Figure 58b and c) are indeed nonplanar, we see that some graphs are planar and others aren't. We might ask, "Why?" Or in expanded form, "What is it about the structure of a graph that makes it planar or nonplanar; is there some specific feature that we can look for?" In this chapter we shall try to answer this question.

Perhaps a reasonable first step toward finding an answer would be to examine a large number of planar and nonplanar graphs, hoping to detect a pattern. There is no shortage of planar graphs. By never allowing edges to cross, you could draw fifty in ten minutes. But genuine (proven) examples of nonplanar graphs are so far nonexistent, since impossibility is nigh impossible to prove. Therefore we shall begin our research by trying to prove that certain graphs which seem to be nonplanar are in fact nonplanar.

UG, K_5, and the Jordan Curve Theorem

We might as well start by proving that Dudeney's puzzle UG is nonplanar. To so do we shall yank a theorem and corollary (a "corollary" is a theorem associated with another theorem from which it can be easily derived) from an advanced level of another branch of mathematics. The theorem is named after the French mathematician Camille Jordan (1838–1922) who first enunciated it in 1899.

Jordan Curve Theorem. If C is a continuous simple closed curve in a plane, then C divides the rest of the plane into two regions having C as their common boundary. If a point P in one of these regions is joined to a point Q in the other by a continuous curve L in the plane, then L intersects C.

Corollary. If C is a continuous simple closed curve in a plane and two points of C are joined by a continuous curve L in the plane having no other points in common with C, then except for its endpoints L is entirely contained in one of the two regions determined by C.

Please don't panic. The theorem and corollary actually say very simple things, though this is obscured by technical jargon. A "continuous curve" is roughly (we haven't the space to do better than "roughly") an unbroken one-dimensional figure, something one could draw with a pencil without lifting the pencil from the paper. Don't be misled by the word "curve"; it is a technical term and pertains to a wide variety of things, including straight lines. For example the five drawings of Figure 59 are continuous curves. A continuous curve is *closed* if the starting point and ending point are the same; thus only c), d), and e) of Figure 59 are continuous closed curves. And a continuous closed curve is *simple* if no point other than the starting point is repeated, and the starting point itself is repeated only once. So only e) of Figure 59 is a continuous simple closed curve. Figure 60 has two more examples of continuous simple closed curves. Notice that "simple" in the mathematical sense is not the same as "simple" in everyday usage—Figure 60b is mathematically "simple".

Roughly then, a continuous simple closed curve is something you can make out of a rubber band without cutting it or folding it back

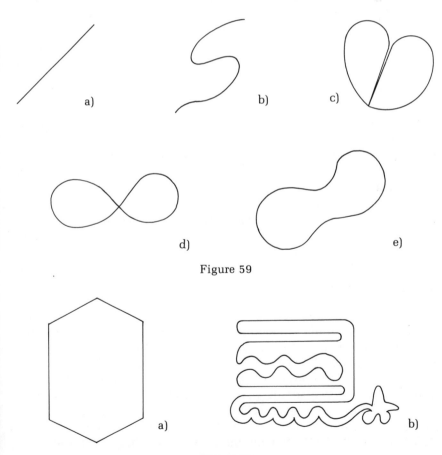

a)   b)   c)

d)   e)

Figure 59

a)   b)

Figure 60

onto itself. The Jordan Curve Theorem says that any such thing embedded in a plane cuts the plane into two regions—the inside and the outside—and that any unbroken curve drawn in the plane from a point inside to a point outside must somewhere cross the rubber band. The corollary says that if two points on the rubber band are joined by an unbroken curve in the plane that doesn't touch the rubber band anywhere else, then that unbroken curve must be either entirely inside the rubber band or entirely outside it.

Once they have been translated into plain English the Jordan Curve Theorem and its corollary are painfully obvious. But to a pure mathematician everything that can possibly be proved, no matter how "obvious", requires proof. Remember Russell's Paradox. Sets were once thought to be "obviously" simple things.

As a matter of fact all known proofs of the Jordan Curve Theorem are quite difficult to follow and probably not one in a hundred professional mathematicians has ever seen such a proof. Jordan himself couldn't prove the Jordan Curve Theorem! A few years after his theorem appeared in print, the "proof" he had published along with it was shown to be invalid. Decades passed before another mathematician finally succeeded in constructing a valid proof.

You may wonder at the problems mathematicians have had with the Jordan Curve Theorem, since what it asserts is so "obvious". It often happens in mathematics that the most "obvious" theorems have the most difficult proofs. Perhaps this is because when working with basics there are few concepts to work with and few previously proved theorems.

At any rate, the Jordan Curve Theorem has been included in this book both as an example of the lengths to which mathematicians are prepared to go in pursuit of "rigor", and as a means for making our proof of the next theorem as valid as possible under present mathematical standards.

Theorem 3. *UG is nonplanar.*

Proof. The utility graph can be represented as in Figure 61a. We shall try to draw it without edge-crossings. Then we shall invoke first the corollary and then the Jordan Curve Theorem itself to show that one edge-crossing, despite our efforts, is unavoidable.

Let **C** denote the object in Figure 61b. Note that **C** is a continuous simple closed curve in the plane. Connect A to Z, B to Y, and C to X by continuous curves lying in the plane and having no other points in common with **C**. We shall denote these curves "$\{A,Z\}$", "$\{B,Y\}$", and "$\{C,X\}$", but for the present we shall leave them out of the diagram.

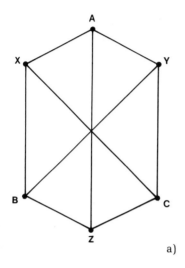

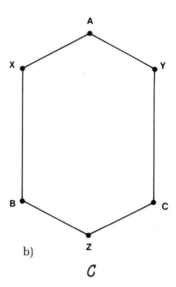

a)   b)

Figure 61

Applying the corollary in turn to **C** and $\{A,Z\}$, **C** and $\{B,Y\}$, and **C** and $\{C,X\}$, we conclude that each of $\{A,Z\}$, $\{B,Y\}$, and $\{C,X\}$ is—except for its endpoints—either entirely inside **C** or entirely outside **C**.

If you had three golf balls and two boxes, and put each ball into one or the other of the boxes, the result would be that exactly one of the boxes would contain two or more balls. In this situation we have curves instead of golf balls and regions instead of boxes, but the so-called "pigeonhole principle" is the same, so we are bound to consider two cases.

Case 1. There are at least two curves inside **C**.

Say that $\{A,Z\}$ and $\{B,Y\}$ are the two curves definitely inside **C**. It's true that we are restricting ourselves to a specific situation, but the following argument would hold for any other two of the three curves.

C with $\{A,Z\}$ inside it has been drawn in Figure 62a. At this point it is in some sense "obvious" that $\{B,Y\}$ cannot also be drawn inside **C** without an edge-crossing. But appeals to intuition are mathematically suspect, so we will add to the picture only what we know for sure, namely that $\{B,Y\}$ leaves B and goes inside **C** and comes from the inside of **C** to end at Y. This has been done in Figure 62b.

Now let Q be a point on the curve BY near enough to B so that there are no edge-crossings between B and Q. Then $AZBXA$ is a

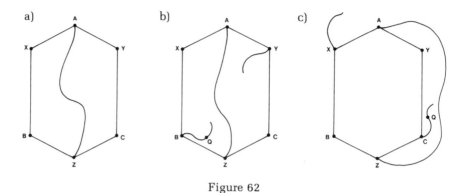

Figure 62

continuous simple closed curve in the plane, Q is a point inside, Y is a point outside, so by the Jordan Curve Theorem any continuous curve in the plane joining Q to Y intersects $AZBXA$. Since $\{B,Y\}$ is such a curve we conclude that $\{B,Y\}$ cannot be drawn without crossing another edge. For this case at least we have proved that UG cannot be drawn in a plane without edge-crossings.

Case 2. There are at least two curves outside **C**.

Say they are $\{A,Z\}$ and $\{C,X\}$. The argument is similar to Case 1. Draw **C** with $\{A,Z\}$ outside it, draw both ends of $\{C,X\}$, and let Q be a point on $\{C,X\}$ near enough to C so that there are no edge-crossings between C and Q. See Figure 62c. $AYCZA$ is a continuous simple closed curve in the plane, Q is inside, X is outside, therefore by the Jordan Curve Theorem every continuous curve in the plane joining Q to X must intersect $AYCZA$. $\{C,X\}$ is such a curve, hence cannot be drawn without crossing another edge. In this case too, UG cannot be drawn in a plane without edge-crossings. As there are no more cases, the theorem is proved.

In some sense UG is the simplest nonplanar graph, as all others have more edges. In another sense K_5 is the simplest nonplanar graph, as all others have more vertices.

Theorem 4. K_5 *is nonplanar.*

Proof. A simple extension of the proof of the last theorem. Let **C** be the object of Figure 63a. **C** is a continuous simple closed curve in a plane. Connect 1 to 3, 1 to 4, 2 to 4, 2 to 5, and 3 to 5 by continuous curves in the plane having no other points in common with **C**. Applying the corollary to **C** and each of these five curves in turn, we see that each of these curves is either, except for its

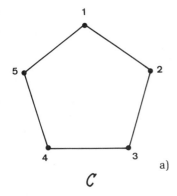

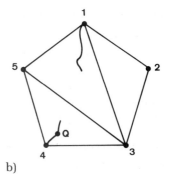

a) b)

C

Figure 63

endpoints, entirely inside **C** or entirely outside **C**.

Invoking the pigeonhole principle, this time with five golf balls and two boxes, we conclude that either the inside of **C** or the outside of **C** contains three or more of the five curves. As before there are two cases to consider.

Case 1. There are at least three curves inside **C**.

Say they are {1,3}, {1,4}, and {3,5}; the following argument would hold as well for any other selection. Note that it is possible to select from the three curves two that share a vertex: Exercise 1 says that this must always happen. And so from the three curves we select two sharing a vertex and draw them inside **C**. Adding the ends of the third curve results in something like Figure 63b.

Let Q be a point on the curve {1,4} near enough to 4 that there are no edge-crossings between Q and 4. 3453 is a continuous simple closed curve in the plane, Q is inside, 1 is outside, so by the Jordan Curve Theorem {1,4}—being as it is a continuous curve in the plane joining Q to 1—must intersect 3453. Thus for this case K_5 cannot be drawn in a plane without edge-crossings.

Case 2. There are at least three curves outside **C**.

I leave the proof of this case to you.

We have invested thus far quite a bit of time and energy, the result being a grand total of two nonplanar graphs. At this rate it will be some time before we have enough specimens for serious research, but fortunately there is a simple theorem we can use to generate infinitely many nonplanar graphs.

Theorem 5. *Any subgraph of a planar graph is planar.*

Proof. Let G be any planar graph. A planar graph can be drawn in a plane without edge-crossings, so let us suppose that this has already been done to G.

A subgraph of G is the result of selective erasing. If no edge or vertex of G is removed the subgraph obtained is G itself, which is planar by hypothesis. Otherwise some edges and/or vertices of G are actually erased. But erasing edges and/or vertices can never create an edge-crossing. So regardless of how much erasing is done, the subgraph obtained will be without edge-crossings and therefore be planar.

Definition 19. If a graph H is a subgraph of a graph G, we will also say that G is a *supergraph* of H.

This definition doesn't present anything new; "G is a supergraph of H" means exactly the same thing as "H is a subgraph of G." Its only purpose is that it will make some of our statements less complicated.

Corollary 5. *Every supergraph of a nonplanar graph is nonplanar.*

Proof. This is just Theorem 5 stated in apparently different but logically equivalent form. Let G be a supergraph of a nonplanar graph H. If G were planar then H, being a subgraph of G, would be planar by Theorem 5. But H is nonplanar, so G must be nonplanar too.

Since a subgraph is the result of selective erasing it follows that a supergraph is the result of selective augmentation. Given a graph G you can form a supergraph of G by adding new vertices and/or connecting nonadjacent (new or old) vertices. But as before there are two catches. First, you needn't add anything, since every graph is a supergraph of itself. Second, you must never put a new vertex on an old edge, for then you are actually replacing that edge by two new ones plus a vertex, so the two formerly adjacent vertices are no longer adjacent, and the resulting graph may not have G as a subgraph. For example the graph of Figure 64a is not a supergraph of Figure 64b; using only an eraser you can never make C_4 look like K_3.

Applying the selective augmentation process to a graph G, it is clear that there is no end of possibilities. Every graph has infinitely many supergraphs. In particular there are infinitely many supergraphs of UG and infinitely many supergraphs of K_5, and all these are guaranteed nonplanar by Corollary 5. Despite the fact that some of

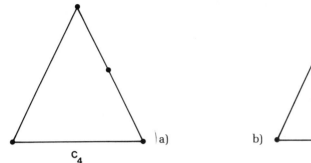
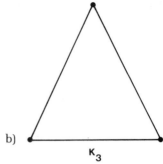

Figure 64

the supergraphs of *UG* are also supergraphs of K_5, and vice versa, we are still in possession of infinitely many nonplanar graphs.

Examples. 1) Figure 65a is nonplanar because it is a supergraph of *UG*.
  2) Figure 65b is nonplanar because it is a supergraph of K_5.
  3) Figure 65c is nonplanar because it is a supergraph of both.
  4) Figure 65d is nonplanar because it is a supergraph of both.
  5) For $v \geq 6$, K_v is nonplanar because it is a supergraph of both.

So here we are with infinitely many nonplanar graphs. We wanted them so that we could examine their structure and perhaps detect

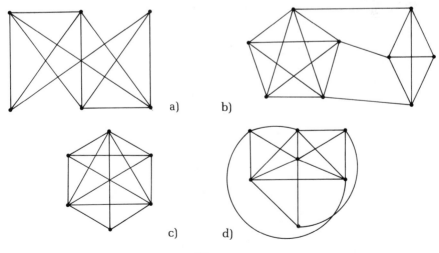

Figure 65

a pattern of some kind. But the only obvious pattern—that they are all supergraphs of UG or K_5—is no help, because that's how we produced these examples in the first place. So it seems that we should start looking around for more nonplanar graphs.

Are there more nonplanar graphs?

At this point every nonplanar graph we know of is a supergraph of UG or K_5. The natural question is, "Are there any more?" If we find some we can compare them to the ones we now have and try to discern properties common to the two groups. Any such properties not shared by planar graphs would be bound up with the inner workings of nonplanarity and we would be that much closer to understanding what makes graphs nonplanar. On the other hand, if a persistent search fails to uncover any new nonplanar graphs, we should suspect that there are no more to be found and should turn our efforts toward proving a theorem that says, "A graph is nonplanar if and only if it is a supergraph of UG or K_5." Proving such a theorem would mean that the only pattern to nonplanar graphs is the one we've already noticed. This would answer the question posed at the beginning of the chapter and, its purpose accomplished, the chapter would contentedly expire.

But a glance at the table of contents shows that this chapter doesn't end for some time, so it seems that indeed there are more nonplanar graphs to be found. It would be instructive if, at the end of this paragraph, you were to put the book away and search for some. Remember the goal: a graph that seems nonplanar—we can worry about proving that later—but is not a supergraph of UG or K_5. Such things do exist, and some are quite simple. There is one with $v = 6$ and $e = 11$, another with $v = 7$ and $e = 10$.

Welcome back. The graphs of Figure 66 are nonplanar, and neither is a supergraph of *UG* or K_5. There are infinitely many other such graphs, some of which you may have discovered, but these two are the simplest, in the following sense. The first has fewer vertices (six) than any other "new type" of nonplanar graph, and the second has fewer edges (ten) than any other "new type" of nonplanar graph.

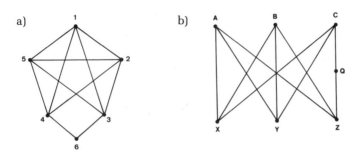

Figure 66

Let us now prove that these are in fact examples of a new type of nonplanar graph.

Consider graph a). It is not a supergraph of *UG* because *UG* has six vertices of degree 3 and so any supergraph of *UG* must contain six vertices with degrees greater than or equal to 3; but this is not true of graph a). Graph a) is not a supergraph of K_5 because K_5 consists of five vertices each adjacent to the other four and so any supergraph of K_5 must also contain five vertices each adjacent to the other four; but the only reasonable choice of five vertices from graph (a)—1, 2, 3, 4, and 5—is missing an edge, for 3 is not adjacent to 4.

All that remains is to show that graph a) is nonplanar. Suppose for the sake of argument that graph a) is planar. Then it can be drawn in a plane without edge-crossings. Suppose that this has been done. Then we can take the crossing-free drawing of graph a) and find vertex 6, which is easy to do since vertex 6 is the only vertex of degree 2. Having found vertex 6 we can erase it and join its two incident edges to one another, forming one new edge where formerly there were two edges and vertex 6.

You might object that this procedure is illegal because we have erased a vertex without also erasing its incident edges, contrary to what we said about erasing on page 32. But that restriction is pertinent only when our goal is a subgraph. We can certainly erase vertex 6 and join the two edges left dangling; it's just that what we end up with will not be a subgraph of graph a).

It is clear that this procedure creates no edge-crossings, and that the graph we have as a result is K_5. Since we started with a plane drawing of graph a) that was crossing-free, the altered drawing is a crossing-free plane drawing of K_5. Of course, no such drawing can possibly exist, as K_5 is nonplanar. This contradiction renders untenable our original supposition that graph a) is planar, so graph a) must be nonplanar.

In a similar way we can show that graph b) is a nonplanar graph, but is not a supergraph of either UG or K_5. It is not a supergraph of UG because UG has six vertices, each of which is joined to three of the other five, and consequently any supergraph of UG must contain a collection of six vertices, each of which is joined to three of the other five; but since Q has degree 2, the only possible collection of six vertices from graph b) is A, B, C, X, Y, Z, and C and Z are joined to only two of the other five. Graph b) is not a supergraph of K_5 because K_5 has five vertices of degree 4 and consequently any supergraph of K_5 must contain five vertices whose degrees are at least 4; but graph b) has only vertices of degrees 2 and 3. Finally, if graph b) were planar we could take a crossing-free drawing of graph b) in a plane, erase vertex Q and join the dangling edges, and thereby create a crossing-free plane drawing of UG, which we know is impossible; so graph b) must be nonplanar.

Expansions

Having seen the two examples of the last section, it seems likely that other similarly-constructed graphs, like those in Figure 67, would also turn out to be nonplanar graphs that are not supergraphs of UG or K_5. So we are motivated to define a new term.

Definition 20. If some new vertices of degree 2 are added to some of the edges of a graph G, the resulting graph H is called an *expansion* of G.

a) 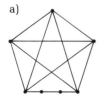 b) 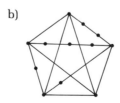 c)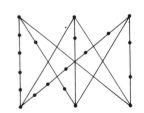

Figure 67

Examples. The graph of Figure 67a is an expansion of K_5; b) is another expansion of K_5; c) is an expansion of UG. The graph of Figure 64a is an expansion of 64b.

In a mathematical context the word "some" includes the possibility "none". By the above definition therefore, every graph is an expansion of itself.

It's easy to confuse "expansion" and "supergraph". As the distinction is crucial to the rest of this chapter, I want to spend some time discussing the two notions.

Both expansions and supergraphs are "augmentations" of a graph, but they are accomplished by different procedures. In making an expansion there is only one thing we are allowed to do (provided we do anything at all): splice new vertices of degree 2 into the edges. That amounts to severing old connections between vertices and inserting intermediary vertices. Though I realize that I can physically "add" vertices of degree 2 to a diagram with only a pencil, I prefer to interpret expansion as something done with pencil and eraser, because this interpretation takes cognizance of the fact that the process of expansion not only augments the structure of a graph, but also subtracts from it.

To make an expansion of a graph,

1) you don't have to do anything (every graph is an expansion of itself);
2) if however you choose to do something, this is what you do: with an eraser, erase some holes in some of the edges, then with a pencil fill the holes with vertices of degree 2.

This is illustrated in Figure 68. The first drawing is the original graph; in the second drawing edges have been selected and holes erased; and the third drawing shows the finished expansion.

As opposed to the process of expansion, in which the only thing we are allowed to do is splice new vertices of degree 2 into the edges, in making a supergraph splicing new vertices (of any degree) into the edges is the only thing we are *not* allowed to do. We mentioned this on page 72 and cited as an example the graphs of Figure 64, C_4 and K_3. In order for C_4 to be a supergraph of K_3, K_3 would (by Definition 19) have to be a subgraph of C_4. And I said on page 72 that "using only an eraser you can never make C_4 look like K_3," so K_3 is not a subgraph of C_4. I was referring to the "selective erasing" process whereby we interpreted subgraph extraction back on page 32. It might be helpful if I repeated it here.

To make a subgraph of a graph,

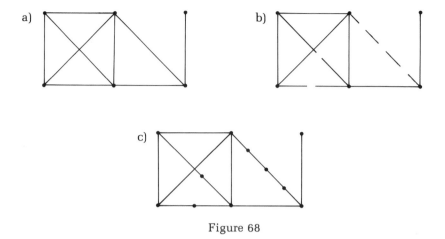

Figure 68

1) You don't have to do anything (every graph is a subgraph of itself);

2) if however you choose to do something, this is what you do: with an eraser erase however many vertices and/or edges as you like, subject to the restriction that when you erase a vertex you must also erase all edges incident to it.

With this interpretation it's clear that K_3 is not a subgraph of C_4 and therefore that C_4 is not a supergraph of K_3. As C_4 can be produced from K_3 by splicing a vertex into one edge, we should prohibit splicing whenever the goal is the construction of a supergraph.

To make a supergraph of a graph,

1) you don't have to do anything (every graph is a supergraph of itself);

2) if however you choose to do something, this is what you do: with a pencil, add as many vertices as you like, provided you don't splice them into existing edges; and add as many edges as you like, anywhere you like—new edges may join two original vertices, or two new vertices, or one original vertex and one new vertex.

With these interpretations we see that the processes of expansion and making a supergraph involve different tools. To make an expansion you use both pencil and eraser, while to make a supergraph you use only a pencil (and to make a subgraph you use only an eraser).

Not only are expansions and supergraphs accomplished by different procedures involving different tools, but they usually have different results as well. That is, an expansion of a graph G is usually not

a) 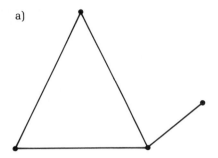   b)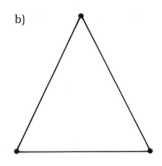

Figure 69

a supergraph of G, and a supergraph of G is usually not an expansion of G. For example we have already seen that C_4 is an expansion of K_3 but not a supergraph of K_3. Figure 69 provides an example of the reverse situation. The first graph is a supergraph of K_3. It was constructed from K_3 using only a pencil, by drawing one new vertex and one new edge. It is not an expansion of K_3 because the new vertex was not spliced into an edge, and besides it is not of degree 2.

I said that an expansion is "usually" not a supergraph, and vice versa. There are exceptions.

Example. In Figure 70, a) is a supergraph of b) from the perspective that 1 corresponds to 3 with vertex 2 and edge {1,2} added on with a pencil. From another perspective a) is an expansion of b): 2 corresponds to 3 with vertex 1 spliced in with eraser and pencil.

Exceptions notwithstanding, it is generally true that expansions and supergraphs are different, and so it is important to keep the notions distinct in your mind.

The graph of Figure 71a is not an expansion of Figure 71b. This

a) 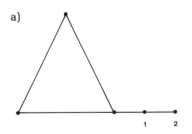   b)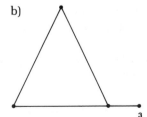

Figure 70

a)
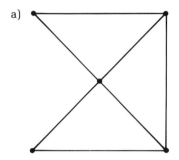

b)
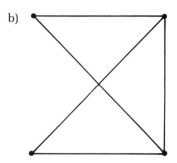

Figure 71

is because Definition 20 insists that spliced vertices have degree 2. So we see that an expansion of a graph has the same number of edge-crossings as the original. It is this property that accounts for the following theorem.

Theorem 6. *Every expansion of UG or K_5 is nonplanar.*

Proof. The proof is a generalization of the argument we used twice in the last section. Let H be any expansion of UG or K_5. Suppose for the sake of argument that H is planar. Then H can be drawn in a plane without edge-crossings. Suppose this has been done.

Now let's examine the crossing-free drawing of H we are supposing to exist. All the "new" vertices are easy to find because they are all of degree 2, whereas the original vertices all have higher degree (3 if H is an expansion of UG; 4 if H is an expansion of K_5). Erase all the new vertices and in each case join the two dangling edges to form a single edge. This reversal of the process that produced H will not create any edge-crossings because the vertices we have removed are all of degree 2. Thus our alteration of the crossing-free drawing of H is also crossing-free. But our alteration is nonplanar, being either UG or K_5. This contradiction shows our supposition to be false, so H is nonplanar.

Corollary 6. *Every supergraph of an expansion of UG or K_5 is nonplanar.*

Proof. Let H be an expansion of UG or K_5 and let G be a supergraph of H. H is nonplanar by Theorem 6, so G is nonplanar by Corollary 5.

Kuratowski's Theorem

We started with the two simplest nonplanar graphs, UG and K_5. We used Corollary 5 to generate infinitely many more, the supergraphs of UG and K_5. Then we used Theorem 6 to generate another infinity of examples, the expansions of UG and K_5. Now to this massive infinity of nonplanar graphs Corollary 6 adds yet another infinitely many, the supergraphs of the expansions of UG and K_5.

The situation is illustrated in Figure 72. The right-hand circle represents the set of all graphs that are supergraphs of UG or K_5. UG and K_5 are themselves contained in this circle because they are supergraphs of themselves. The left-hand circle represents the set of all graphs that are expansions of UG or K_5. UG and K_5 are contained in this circle as well because they are expansions of themselves. Otherwise the two circles have nothing in common, because UG is the only expansion of UG that is also a supergraph of UG, and K_5 is the only expansion of K_5 that is also a supergraph of K_5 (see Exercise 13).

SUPERGRAPHS OF EXPANSIONS

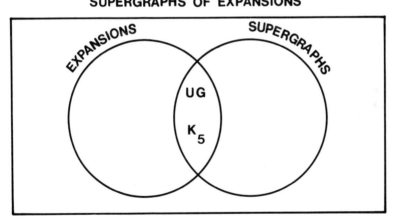

Figure 72

The rectangle represents the set of all graphs that are supergraphs of expansions of UG or K_5. It contains everything in the two circles, and infinitely many other things besides. It contains the left-hand circle because an expansion of (say) UG can be viewed as a supergraph of itself and therefore as a supergraph of an expansion of UG. The rectangle contains the right-hand circle because UG can be viewed as an expansion of *UG* and so a supergraph of *UG* can be viewed as a supergraph of an expansion of *UG*. And the graph of Figure 73 is an example of something contained in the rectangle but not contained in either circle.

The graph of Figure 73 is not an expansion of *UG* because expansions of *UG* have six vertices of degree 3 (the original vertices) and have all other vertices of degree 2 (the new vertices spliced into the edges); but Figure 73 has a different distribution of degrees. And it's not a supergraph of *UG* because *UG* consists of six vertices each of which is joined to three of the other five, so any supergraph of *UG* would have to contain a collection of six vertices each of which is joined to three of the other five; but since the degrees of vertices 1, 2, 3, and 4 are too small, the only possible collection of six vertices from Figure 73 is *A, B, C, X, Y, Z*, and of these *A, X, C,* and *Z* are only joined to two of the other five. Clearly the graph of Figure 73 is neither an expansion nor a supergraph of K_5. It is actually a supergraph of an expansion of *UG*. It was formed by first splicing three vertices into the edges of *UG*, creating the expansion of *UG* drawn in Figure 74, and then making a supergraph of Figure 74 by adding a vertex and two edges. The result (Figure 73) is nonplanar by Corollary 6.

The first nonplanar graphs we discovered were the supergraphs of *UG* and K_5. Then we discovered a second type, the expansions of *UG* and K_5. The graph of Figure 73 is our first specific example of a third type that includes the other two. Since this third type has resulted from taking supergraphs of expansions of *UG* or K_5, you might wonder if taking expansions of supergraphs of *UG* of K_5 would result in a fourth type. As a matter of fact, it wouldn't. Every expansion of a supergraph of *UG* or K_5 is also a supergraph of an

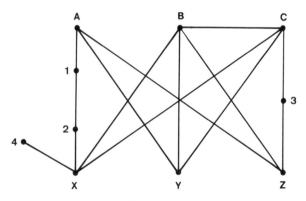

Figure 73

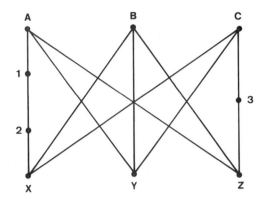

Figure 74

expansion of *UG* or K_5. For example, Figure 75 traces the construction of an expansion of a supergraph of K_5, but Figure 76 shows how the same result can be achieved by taking a supergraph of an expansion of K_5.

If there are still more nonplanar graphs floating around we should try to discover them and compare them to the ones we have now in order to understand what makes these things tick. On the other hand, if a persistent search reveals no additional types of nonplanar graph, we should try to prove a theorem that says there are no more nonplanar graphs to be found. Such a theorem would mean that in constructing the examples we now have, we have already discovered the pattern to nonplanar graphs.

Once more we are faced with the question, "Are there more nonplanar graphs?" That is, "Are there any nonplanar graphs which are not supergraphs of expansions of *UG* or K_5?"

In the early decades of this century mathematicians asked this question, searched for new kinds of nonplanar graphs, and failed to find any. So they took the next step and tried to prove that there

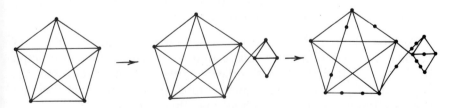

Figure 75

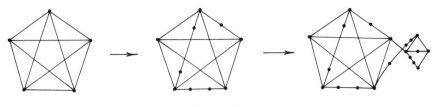

Figure 76

were none to be found. In 1930 a mathematician named Kasimir Kuratowski succeeded in doing this.

Kuratowski's Theorem. *Every nonplanar graph is a supergraph of an expansion of UG or K_5.*

The proof is rather involved, and I will omit it. It may seem that in citing Kuratowski's Theorem without proof I am pulling a rabbit out of a hat, much as I did with the Jordan Curve Theorem. The situations, however, are not parallel. I *couldn't* prove the Jordan Curve Theorem without another book. I *could* prove Kuratowski's Theorem in considerably less space, maybe thirty pages, but I have decided not to because it doesn't seem worth it. Apart from being long, the proof is highly specialized and has almost no bearing on the other topics we will be taking up. Suffice it to say that Kuratowski's Theorem has been proved, and that if you wish to see a proof you can find one in Harary's *Graph Theory*, pp. 108–113. (At first you may find Harary's proof difficult because it is rather condensed. But if you are willing to spend the time necessary to look up and absorb all unfamiliar concepts and symbols, you should be able to follow it well enough.)

Corollary 7. *The set of all nonplanar graphs is equal to the set of all graphs that are supergraphs of expansions of UG or K_5.*

Proof. Let N be the set of all nonplanar graphs and E be the set of all graphs that are supergraphs of expansions of UG or K_5. N is a subset of E by Kuratowski's Theorem and E is a subset of N by Corollary 6, so $N = E$ by Definition 4.

Corollary 7 is quite a surprise. It says that beneath the apparent chaos of nonplanar graphs there is an order of astounding simplicity. By it we see that the two simplest nonplanar graphs, UG and K_5, are essentially the *only* nonplanar graphs, in the sense that they are the building-blocks with which all the others are constructed.

Our original question has been answered. We have penetrated the nature of nonplanarity and have been able to explain it in terms of other concepts that are more easily understandable. It follows that we have also penetrated the nature of planarity. By Corollary 7, "nonplanar" means the same as "being a supergraph of an expansion of UG or K_5," so "planar" must mean "not being a supergraph of an expansion of UG or K_5." It is an intellectual delight to find such a simple solution to such a complex problem.

Corollary 7a. *The set of all planar graphs is equal to the set of all graphs that are not supergraphs of expansions of UG or K_5.*

Determining whether a graph is planar or nonplanar

At the beginning of this chapter, proving a graph to be nonplanar consisted of proving that it could not possibly be drawn in a plane without edge-crossings, a negative procedure that for even the simplest nonplanar graphs was rather tedious (Theorems 3 and 4). Corollary 5 gave us a shorter and more positive method: if we can find a subgraph of a graph G that is isomorphic to UG or K_5, then that proves that G is nonplanar. Unfortunately we had no reason to believe this method would always work. And we soon found that indeed it would *not* always work: the graph of Figure 66a is nonplanar but contains no subgraph isomorphic to UG or K_5. This discovery led to Theorem 6, which in conjunction with Corollary 5 gave us an improved method for proving a graph nonplanar: if, given a graph G, we can either find a subgraph isomorphic to UG or K_5, or we can show that G itself is isomorphic to an expansion of UG or K_5, then that proves that G is nonplanar. Again we had no reason to believe this method would always work, and again it turned out that sometimes it would not: the graph of Figure 73 is nonplanar, but is neither a supergraph nor an expansion of UG or K_5. But by then we had formulated Corollary 6, which gave us a third method: if we can find a subgraph of a graph G that is isomorphic to an expansion of UG or K_5, then that proves G is nonplanar. The significance of Kuratowski's Theorem is that by saying every nonplanar graph contains such a subgraph, *it guarantees that the third method will always work.*

Often we will encounter a graph that we want to categorize as being either planar or nonplanar, but which is not obviously either. Here are some guidelines for determining which it is.

1) Before you do anything else, try to draw the graph without edge-crossings. If you succeed that proves it's planar and the matter

is settled. If you fail then you know it's probably nonplanar.

2) If you decide the graph is probably nonplanar, the next step is to prove it, which by Corollary 7 amounts to finding a subgraph that is an expansion of *UG* or K_5. The vast majority of nonplanar graphs contain an expansion of *UG*, so start by looking for that. If, however, the graph does not have at least six vertices with degrees greater than or equal to 3, then it cannot possibly contain an expansion of *UG* and you should go to step 4).

3) Here's how to look for an expansion of *UG*. I don't guarantee that this procedure will unfailingly uncover an expansion of *UG*, as there are a number of choices involved; nevertheless I think it is useful as a search pattern. Pick a vertex whose degree is at least 3, and call it the "first house". Then find three other vertices with degrees at least 3 (the "utility companies"), from which it is only a short trip along the edges to the first house. Ideally the three utility companies would be joined to the first house by a single edge, but if this is not possible pick them so that you can travel to the first house by a short alternating sequence of edges and vertices. No two of these paths should share a vertex, other than the vertex you have called the "first house", nor should they share an edge, otherwise the subgraph you are constructing will not untangle into an expansion of *UG*. Mark the paths from the utilities to the first house with jagged lines. Now pick the "second house", a vertex with degree at least 3 from which you can travel to the three utilities along paths that are, except for their endpoints, disjoint from one another and from the three paths going to the first house. Mark these paths. In a similar fashion pick the "third house" and mark three paths joining it to the three utilities; as before these paths should be, except for their endpoints, disjoint from each other and from the previous six paths. Now extract a subgraph from the original graph by erasing everything that you haven't marked. It is easy to show that such a subgraph is isomorphic to an expansion of *UG*. The three houses correspond to vertices *A*, *B*, and *C* of *UG* (as originally defined), the three utilities correspond to vertices *X*, *Y*, and *Z*, and any other vertices of the subgraph correspond to expansion vertices spliced into the edges of *UG*.

4) Here's how to look for an expansion of K_5. (Of course, the graph must have at least five vertices with degrees greater than or equal to 4, or it cannot possibly contain an expansion of K_5.) Pick a vertex (I'll call it "#1") having a degree of at least 4 and mark four paths joining it to four other vertices ("#2", "#3", "#4", and "#5") having degrees that are at least 4. Then from vertex #2 mark three paths

leading to vertices #3, #4, and #5; from vertex #3 mark two paths
leading to vertices #4 and #5; and from vertex #4 mark a path leading
to vertex #5. These ten paths must be mutually disjoint; if they have
any edges in common, or any vertices in common other than their
endpoints, the subgraph you have marked will not be an expansion
of K_5. Extract this subgraph by erasing everything unmarked. It is
a simple matter to show that it is isomorphic to an expansion of
K_5.

Examples. 1) I want to know whether or not the graph in Figure
77 is planar. I start by trying to draw it without edge-crossings. In
my first attempt I pull several of the edges to the outside (Figure
78a). This reduces the number of crossings to four, and I notice that
I can eliminate three of them by bending edge {2,7} and flipping
edge {7,8} end over end (Figure 78b). Then after staring at Figure
78b for a while I notice that I could eliminate the only remaining
crossing by moving edge {5,9}. The result, shown in Figure 79a, is

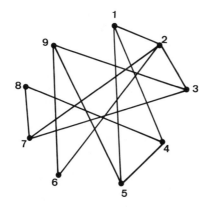

Figure 77

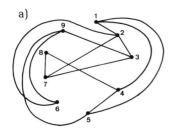
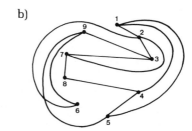

Figure 78

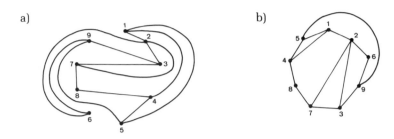

Figure 79

crossing-free and so the original graph, being isomorphic to the graph of Figure 79a, is planar. Figure 79b is a less amorphous-looking version of Figure 79a.

2) The vertices of a graph H correspond to the squares of a 4 × 4 chessboard, and two vertices are joined by an edge whenever a knight can go from one of the corresponding squares to the other in one move. I have drawn H in Figure 80 (messy, isn't it?) and I want to know whether or not it is planar. I start by trying to draw it without edge-crossings. I fail repeatedly; Figure 81 is the best I can do. So now I'm virtually certain that H is nonplanar. I notice that H doesn't contain an expansion of K_5, because it has no vertices of degree 4, so I conclude that H must contain an expansion of UG. (In any case, I would have searched first for an expansion of UG, because expansions of UG are more likely and easier to find.) H has eight vertices of degree 3 and four of degree 4, so I have twelve

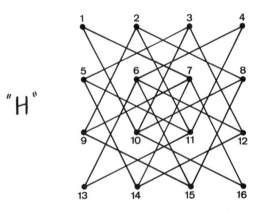

Figure 80

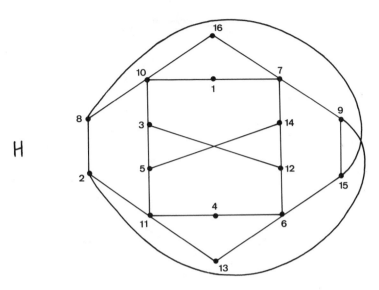

Figure 81

vertices from which to choose my houses and utilities. I'll choose
vertex 8 at random to be the first house. 8 is joined directly to vertices
2, 10, and 15, and these vertices have degrees greater than or equal
to 3, so I'll let them be the three utilities. I'll mark the edges from
8 to 2, 8 to 10, and 8 to 15 (Figure 82). Now I have to pick a second
house. I'll look for a vertex from which I can travel easily to the
three utilities. Vertex 9 seems a likely candidate because it is joined
to two of the utilities directly; and I can get from 9 to the third
by passing through 7 and 16. So I'll make 9 the second house and
mark the paths joining 9 to 2, 9 to 10, and 9 to 15. Finally I need
a third house. Almost any vertex with degree at least 3 would do,
but I'll pick vertex 3 (because I feel like it) and mark three paths
joining 3 to the utilities; the path from 3 to 10 is a single edge,
the path from 3 to 2 passes through 5 and 11, and the path from
3 to 15 passes through 12 and 6. This is all displayed in Figure
82. Except for the houses and utilities no vertex has been marked
more than once, and no edge has been marked more than once, so
I know the paths are disjoint and therefore that I'm practically finished.
I'll erase every unmarked vertex and edge, extracting from H the
subgraph J drawn in Figure 83. In Figure 84a I draw an unlabeled
version of UG and give its houses (the vertices in the top row) the
same labels as the houses in J, and its utilities (the vertices in the

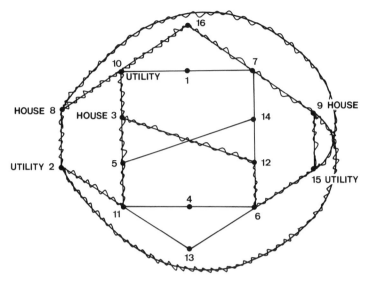

Figure 82

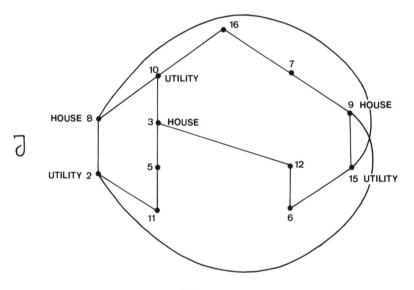

Figure 83

bottom row) the same labels as the utilities in *J*. Then I expand *UG* in accordance with the paths in *J*; for instance, in graph *J* house 9 is joined to utility 10 via 7 and 16, so into the edge {9,10} of *UG* I'll splice two vertices of degree 2 and label them 7 and 16.

a)
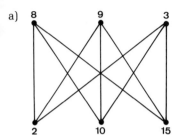

b)
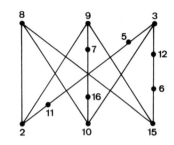

Figure 84

Continuing the process I arrive at the expansion of *UG* drawn in Figure 84b. The graph in 84b is isomorphic to *J*, which in turn is a subgraph of *H*, so I have shown that *H* is a supergraph of an expansion of *UG* and therefore is nonplanar by either Corollary 6 or Corollary 7.

3) The graph of Figure 85 is both a supergraph of an expansion of *UG* and a supergraph of an expansion of K_5. For the sake of example I'll extract an expansion of K_5. The first thing I notice is that the graph is nearly symmetric, so it probably won't matter which five vertices I select to call "#1", "#2", "#3", "#4", and "#5". But to be safe I'll make sure I include the two (*A* and *D*) having a higher degree than the others. So without further ado I'll let *G* be "#1", *A* be "#2", *B* be "#3", *C* be "#4", and *D* be "#5". #1 is joined directly to #2 and #3, so I'll mark these two edges with a jagged line; I'll return later to devise paths from #1 to #4 and #5. #2 is joined directly to #3, #4, and #5, so I'll mark these edges; #3 is joined directly to #4 and #5, so I'll mark these edges; and #4 is

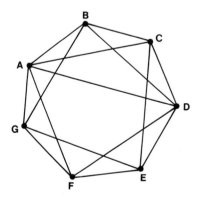

Figure 85

joined directly to #5, so I'll mark that edge. Now I can see what
material is left for the construction of paths from #1 to #4 and from
#1 to #5. I see that I can go from #1 to #4 via E, so I'll mark

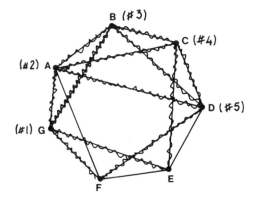

Figure 86

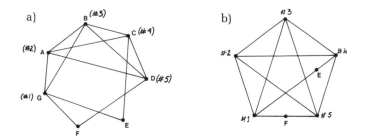

Figure 87

that path. I can't go from #1 to #5 via F and E, for such a path
would have vertex E in common with the last path, and the marked
subgraph would not be an expansion of K_5; but I see another way
of going from #1 to #5, a way that uses only F, so I'll choose and
mark that path. The marked paths are displayed in Figure 86. Now
I'll erase everything I haven't marked, leaving the subgraph drawn
in Figure 87a. This subgraph is isomorphic to the graph of Figure
87b; as this last is obviously an expansion of K_5, I'm finished.

Exercises

1. Prove the following statements:
   a) If three edges are added to the graph of Figure 63a, then at

least two of the new edges will be adjacent.

b) Every graph with $v = 5$ and $e = 3$ has at least two adjacent edges.

c) If v is an odd number then every graph with v vertices and $(1/2)(v + 1)$ edges has at least two adjacent edges.

2. Prove that in Figure 58, graph a) is planar, and that graphs b) and c) are nonplanar.

3. The expansions of K_3 are the cyclic graphs C_3, C_4, C_5, etc. Satisfy yourself that except for C_3 (which is isomorphic to K_3), no cyclic graph is a supergraph of K_3. Thus K_3 has the property that none of its expansions (except itself) is also a supergraph. Find a simple graph having the "reverse" property; that is, find a graph G such that *every* expansion of G is also a supergraph of G.

4. Find a planar graph with $v = 8$ whose complement is also planar.

5. It is a fact that every planar graph with $v = 9$ has a nonplanar complement. Verify this in one case by drawing a planar graph with $v = 9$ (make it complicated or the exercise will be boring) and proving that its complement is nonplanar.

6. Draw a nonplanar graph whose complement is nonplanar.

7. Prove that the "Petersen graph" of Figure 88 is nonplanar.

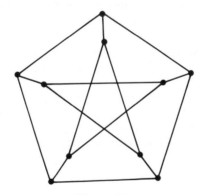

Figure 88

8. Of the 34 graphs with $v = 5$, K_5 is of course the only one that is nonplanar. But of the 156 graphs with $v = 6$, 13 are nonplanar besides *UG*. Find them.

9. Prove: if H is an expansion of G then $v_G + e_H = v_H + e_G$ (where v_G, e_H, v_H, and e_G denote respectively the number of vertices of G, the number of edges of H, the number of vertices of H, and the number of edges of G).

10. Prove: if H and J are expansions of G then $v_H + e_J = v_J + e_H$ (where

as above the subscripts tell us the graph whose vertices or edges are being counted).

11. Find all integers v for which $\bar{C}_v$ (the complement of C_v) is nonplanar. Prove that your answer is correct.

12. Here is an explicit statement of the "pigenonhole principle" mentioned in the text on pages 69 and 71:

    If m objects are distributed into n boxes and m is larger than n, then at least one box contains m/n of the objects.

    Use this principle to prove that there are at least two red maples in the United States having the same number of leaves.

13. Prove the following statements:

    a) Except for UG itself, no expansion of UG is also a supergraph of UG.

    b) Except for K_5 itself, no expansion of K_5 is also a supergraph of K_5.

14. Let S be the set of all expansions of supergraphs of UG or K_5, and let T be the set of all supergraphs of expansions of UG or K_5. We mentioned on pages 82–83 that every expansion of a supergraph of UG or K_5 is also a supergraph of an expansion of UG or K_5, so S is a subset of T. Prove that on the other hand T is *not* a subset of S, and therefore $S \neq T$, by finding a supergraph of an expansion of K_5 that is not also an expansion of a supergraph of K_5.

15. Isomorphism is "transitive", that is, if $G \cong H$ and $H \cong J$, then $G \cong J$. This enables us to prove the following theorem.

    Theorem. *If H is planar and G $\cong$ H, then G is planar also.*

    Proof. H is planar, so H is isomorphic to a graph J which has been drawn in a plane without edge-crossings (Definition 18). Since $G \cong H$ and $H \cong J$, $G \cong J$; that is, G is isomorphic to a graph which has been drawn in a plane without edge-crossings. Therefore G is planar.

    Thus planarity is yet another property preserved by isomorphism (the seventh we've mentioned). Use this fact to devise new proofs that the pairs of graphs in Figures 46, 51, 54, and 55 are not isomorphic.

16. Prove that the graphs in Figure 89 are planar.

17. Prove that the graphs in Figure 90 are nonplanar.

18–20. In Figures 91–93, decide whether each graph is planar or nonplanar and then prove that your choice is correct.

a)   b)

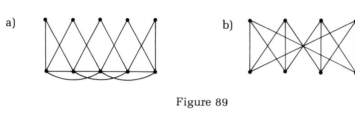

Figure 89

a)   b)

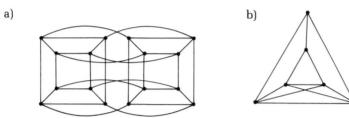

Figure 90

a)   b)

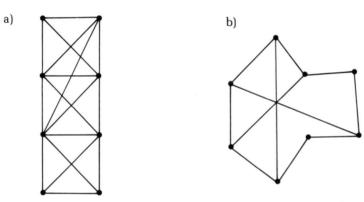

Figure 91

a)   b)

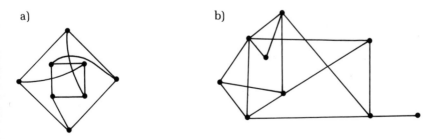

Figure 92

a) 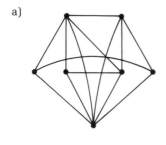   b)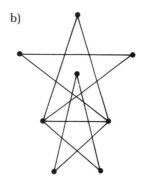

Figure 93

Suggested reading

On Mathematical Jargon

*"New Names for Old" by Edward Kasner and James R. Newman, reprinted in volume 3 of *The World of Mathematics* edited by James R. Newman (Simon and Schuster, 1956).

On Impossibility

The Borders of Mathematics by Willy Ley (Pyramid Publications, 1967).

4. EULER'S FORMULA

Introduction

Leonhard Euler (pronounced "oiler", 1707–1783) is judged by all to have been the most productive, and by many to have been the best, mathematician of modern times. He was Swiss, but spent much of his life in Russia because he had a big family and Catherine the Great offered him a lot of money. His paper "The Seven Bridges of Königsberg" (1736), which we will discuss in Chapter 8, is the earliest known work on the theory of graphs.

The theorem now known as Euler's Formula was proved by Euler in 1752. It is one of the classic theorems of elementary mathematics and plays a central role in the next three chapters of this book. To state it we need some preliminary definitions.

Definition 21. A *walk* in a graph is a sequence $A_1 \, A_2 \, A_3 \ldots A_n$ of not necessarily distinct vertices in which A_1 is joined by an edge to A_2, A_2 is joined by an edge to A_3, ..., and A_{n-1} is joined by an edge to A_n. The walk $A_1 A_2 A_3 \ldots A_n$ is said to *join* A_1 and A_n.

Examples. 1) Let G be the graph of Figure 94a. Then $DCAB$, $ACDCACD$, $ABABAB$, ACA, and $BACDCBCD$ are all walks in G. $ACDAB$ is not a walk because the third vertex in the sequence is not adjacent to the fourth.

2) Any sequence of distinct vertices in K_v is a walk.

3) A null graph N_v has no walks.

Definition 22. A graph is said to be *connected* if every pair of vertices is joined by a walk. Otherwise a graph is said to be *disconnected*.

Examples. 1) The graphs of Figure 94a and 94b are connected.

2) The graphs of Figure 95a and 95b are disconnected, for in each case it is possible to find a pair of vertices not joined by any walk.

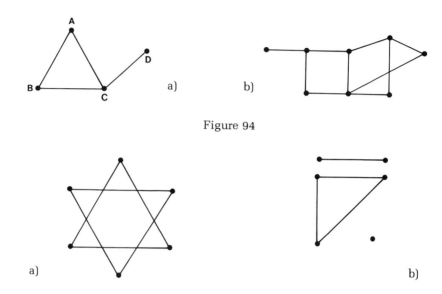

Figure 94

Figure 95

3) Every cyclic graph is connected, as is every complete graph.

4) Except for N_1, all null graphs are disconnected.

You may be puzzled at my statement that N_1 (also known as K_1) is connected, as the definition refers to *pairs* of vertices. The reason is that the system of logic used in the mainstream of mathematics contains the Law of Excluded Middle, whereby every properly formed statement is either true or false and there is no third possibility. In everyday life this is not always the case; for example I don't think that the statement "George Washington stopped beating his wife in March 1790" fits very well into either category. But in mathematics there are only the two categories, so being "true" is the same as being "not false". It follows that N_1 is connected, and we might argue as follows.

In order for N_1 to be disconnected it would, by Definition 22, have to contain two vertices which are not joined by any walk. But N_1 doesn't even contain two vertices, let alone two not joined by a walk. So the statement "N_1 is connected" is not false and must therefore be true.

Intuitively, connected graphs are in one piece and disconnected graphs aren't. The graph of Figure 95a looks at first as if it were in one piece, but it really isn't because it is isomorphic to a graph formed by placing two copies of K_3 next to one another.

Definition 23. When a planar graph is actually drawn in a plane without edge-crossings, it cuts the plane into regions called *faces* of the graph. The letter "*f*" shall denote the number of faces of a planar graph.

Implicit in this definition are two assumptions. One is that a planar graph, when drawn in a plane without edge-crossings, does indeed cut the plane into well-defined regions (a generalization of the Jordan Curve Theorem). The other assumption is that the number of regions is independent of the particular drawing. Without the first assumption it would be impossible to speak of "faces" at all. And without the second it would be impossible to speak of "the" (unique) number of faces of a specific planar graph. Neither need be assumed, for they can be proved. We will not do so, but please note that without at least knowing that such proofs exist, the definition could not have been validly made.

Examples. In each of the graphs in Figure 96 the faces have been numbered. The first graph has $f = 7$ and the second $f = 10$. Note that the exterior region is counted as a face; it is sometimes called "the infinite face".

Two words of caution. First, only planar graphs have faces; the word "face" is meaningless if used in reference to graph a) in Figure 97, because graph a) is nonplanar. Second, though all planar graphs do have faces, Definition 23 defines them only in connection with *crossing-free drawings* of the planar graphs; K_4 does not have five faces, though it might appear that way in Figure 97b, because 97b

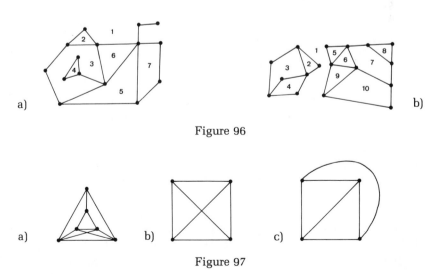

Figure 96

Figure 97

is not crossing-free (actually K_4 has $f = 4$ as you can see from Figure 97c).

The next definition is more convenient than necessary. We will use it to break the proof of Euler's Formula into two pieces, thereby making the proof easier to follow.

Definition 24. A graph is *polygonal* if it is planar, connected, and has the property that every edge borders on two different faces.

Clearly an edge can border on no more than two faces, so what we are excluding are planar connected graphs having one or more edges that border on only one face.

Examples. 1) The graph of Figure 96a is not polygonal because it has two edges that border only on face 1 and another edge that borders only on face 3.

2) Every edge of Figure 96b borders on two faces, but it is not polygonal because it is disconnected.

3) No nonplanar graph has faces, so no nonplanar graph is polygonal.

4) The graphs of Figure 98 are polygonal. Graph b) would have to be redrawn without edge-crossings in order for its polygonal nature to become apparent.

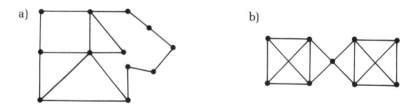

Figure 98

Early mathematicians were amazed to discover that for any circle whatever, no matter how small or large, the circumference C divided by the diameter d always has the same value, 3.14159 .... Of course they couldn't measure the quotient very accurately, but they did know that it was somewhere between $3\frac{1}{8}$ and $3\frac{1}{7}$. In modern terms their discovery was that for circles C/d is constant, and the constant is 3.14159 ....

Euler's discovery was similar. Euler's Formula says that for any planar connected graph whatever, no matter how simple or complicated, the number v of vertices plus the number f of faces minus the number e of edges always has the same value, 2. Or, in modern terms,

Euler's Formula says that for planar connected graphs $v + f - e$ is constant, and the constant is 2. You might want to pause at this point and draw a few dozen planar connected graphs and see for yourself that this is true.

Euler's Formula is rather startling! No matter what graph you draw, provided that it is crossing-free and in one piece, there is the basic simplicity of $v + f - e = 2$.

At the risk of seeming hysterical, we might reflect on how closely Euler's Formula approaches being a fundamental law of nature. Take any situation in which lines cross in a plane—a slice of honeycomb, perhaps, or the paths of gamma-rays on a photographic plate, or a madman's doodle. Count the number of line-crossings and line-ends and call that "v". Count the number of segments into which the lines are cut by the crossings and call that "e". Count the number of regions into which the lines divide the plane and let that be "f". Then provided only that the situation is connected, you are guaranteed the underlying order of $v + f - e = 2$.

Mathematical induction

We shall prove Euler's Formula first for polygonal graphs, and then for planar connected graphs that are not polygonal. All the work is in the first proof. To a mathematician even the first proof is quite simple, employing as it does the familiar (to him/her) technique of "mathematical induction". This section is dedicated to those who have never seen this method of proof.

Principle of mathematical induction. Let S be a statement about positive integers. If one can prove that
    (1) S is true for the positive integer 1, and
    (2) whenever S is true for a positive integer it is true for the next positive integer,
then S is true for every positive integer.

The foregoing is called a "principle" because it is not usually proved, but taken as an axiom for the system of positive integers. A little thought impels us to grant the principle. If we know that statements (1) and (2) are both proved, we can combine them to conclude that the statement S is true for the positive integer 2; then we can combine this new information with statement (2) and conclude that S is true for 3; this in combination with (2) implies that S is true for 4; this and (2) imply that S is true for 5, etc.

A slightly more picturesque model of the principle runs as follows. Take a box of infinitely many dominoes and number them with the positive integers. We will assume that the laws of nature are suspended and that you manage to come up with both the paint and time required to do this. Stand the dominoes on end, in order, on a table top (infinite table required here). We will interpret "S is true for the positive integer n" as meaning that the domino bearing the number n falls over. Then statement (1) of the principle becomes "domino 1 falls over" and statement (2) becomes "whenever a domino falls over the next domino falls over." Under this interpretation the conclusion of the principle says merely that all the dominoes fall over.

The principle is called "induction" because, though it is in fact a method of deductive reasoning, it somewhat resembles the examination-of-cases characteristic of inductive reasoning. If an argument by "mathematical induction" were really inductive, it would have no validity in mathematics.

Proof of Euler's Formula

Theorem 8. *If G is polygonal then v + f − e = 2.*

Proof. The proof is by induction on f. Let S be the statement, "$v + f - e = 2$ holds for all polygonal graphs having f faces". Then S is a statement about positive integers f, and we shall use the principle of mathematical induction to prove that S is true for all positive integers. This of course will mean that the theorem is true.

To begin, then. By the principle of mathematical induction we have to demonstrate two things, the first of which is that

(1) S is true for $f = 1$.

That is, we have to show that $v + f - e = 2$ holds for all polygonal graphs having 1 face. But this is easy. A little thought will reveal that N_1 is the only polygonal graph having just one face (see Exercise 1), and for N_1, $v + f - e = 1 + 1 - 0 = 2$ as required.

The second and final thing we have to demonstrate is that

(2)   if S is true for $f = k$, then S is true for $f = k + 1$.

Translated, we have to show that if $v + f - e = 2$ holds for all polygonal graphs having k faces, then $v + f - e = 2$ holds for all polygonal graphs having $k + 1$ faces. Any "if ...then ..." statement is proved by accepting the first part and deducing the second part. This is no exception, so we will take as given the statement "$v + f - e = 2$ holds for all polygonal graphs having k faces."

Now let G be an arbitrary polygonal graph having $k + 1$ faces. Remove some of the edges and vertices bordering the infinite face of G to produce a new polygonal graph H having one less face than G, so H has k faces. For example if G happened to look like Figure 99a, H might look like Figure 99b.

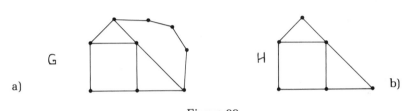

a)   b)

Figure 99

H is a polygonal graph with k faces and we are given that $v + f - e = 2$ holds for all such graphs, so $v_H + f_H - e_H = 2$, where the subscripts help us keep track of the graph whose vertices, faces, and edges we are counting. Our goal is to show that $v_G + f_G - e_G = 2$, so let us try to relate v_G, f_G, and e_G to the corresponding numbers for H.

G has one more face than H because that's where H came from in the first place; thus f_G and f_H have the simple relationship $f_G = f_H + 1$. G has more edges than H, but we don't know how many more. There's a long-standing tradition in mathematics wherein an unknown quantity is represented by "x", so we'll go along with it and let $x = e_G - e_H$. Then e_G and e_H have the relationship $e_G = e_H + x$.

Now think. Whenever edges and vertices are removed from a polygonal graph in order to produce another polygonal graph having one less face, the number of vertices removed is one less than the number of edges removed. For example in Figure 99, four edges and three vertices were removed. Do examples with pencil and paper until you're convinced. Thus $v_G - v_H = x - 1$, that is, $v_G = v_H + x - 1$.

Having these relationships we can evaluate $v_G + f_G - e_G$ by substitution. The following string of equalities shows that when this is done the result is 2.

$$
\begin{aligned}
v_G + f_G - e_G &= (v_H + x - 1) + (f_H + 1) - (e_H + x) \\
&= v_H + x - 1 + f_H + 1 - e_H - x \\
&= v_H + f_H - e_H \\
&= 2.
\end{aligned}
$$

This completes the proof of statement (2). By the principle of mathematical induction statement S is true for every positive integer, and therefore we have the theorem.

Theorem 9. *If G is planar and connected, but not polygonal, then $v + f - e = 2$.*

Proof. Since G is not polygonal it has edges that border on only one face. Let the number of such edges be n. Add one vertex and two edges at each such location, producing a new graph H that is polygonal. For example if G were the graph of Figure 100a then H would look like Figure 100b.

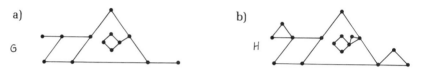

Figure 100

As before we shall compare the numbers of vertices, faces, and edges of the two graphs. Clearly the relationships are $v_G = v_H - n$, $f_G = f_H - n$, and $e_G = e_H - 2n$. Hence

$$
\begin{aligned}
v_G + f_G - e_G &= (v_H - n) + (f_H - n) - (e_H - 2n) \\
&= v_H - n + f_H - n - e_H + 2n \\
&= v_H + f_H - e_H
\end{aligned}
$$

which is equal to 2 by the last theorem.

Euler's Formula. *If G is planar and connected, then $v + f - e = 2$.*

Proof. Combine the two previous theorems.

Euler's Formula is everything a theorem should be. It has the same simple profundity of Kuratowski's Theorem, and an elegant proof—the proofs of Theorems 8 and 9—as well.

Some consequences of Euler's Formula

Theorem 11. *If G is planar and connected with $v \geq 3$, then $(3/2)f \leq e \leq 3v - 6$.*

Proof. Let G be planar and connected with three or more vertices.

Case 1. *G* has a face bounded by fewer than three edges.

Then a little reflection will reveal that *G* must be the path graph P_3 shown in Figure 101a. For P_3, $v = 3$, $f = 1$, and $e = 2$. Hence $(3/2)f = 3/2$, $e = 2$, $3v - 6 = 3$ and the theorem holds.

a)   b)   c)

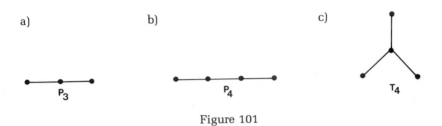

P_3 P_4 T_4

Figure 101

Case 2. Every face of *G* is bounded by three or more edges.

Then numbering the faces of *G* from 1 to *f* we can make the series of statements

$3 \leq$ the number of edges bounding face 1
$3 \leq$ the number of edges bounding face 2
$3 \leq$ the number of edges bounding face 3
$$\vdots$$
$3 \leq$ the number of edges bounding face *f*.

Hence $3f$, the sum of the first column, is less than or equal to the sum of the second column, which we can denote by "*D*". In shorthand, $3f \leq D$. If *G* were polygonal *D* would be equal to $2e$ because every edge of a polygonal graph borders on exactly two faces and so each edge would have been counted exactly twice in the second column, once for each of the two faces it borders. If *G* were not polygonal *D* would be smaller than $2e$, because planar-connected-nonpolygonal graphs contain one or more edges that border on only one face, and such edges would have been counted only once in the second column. So if *G* were polygonal we would have $D = 2e$, whereas if *G* were not we would have $D < 2e$. In either case we can safely say that $D \leq 2e$. Combining this with $3f \leq D$ we have $3f \leq 2e$ which yields the first half of the theorem when both sides are divided by 2.

To get the other half take $(3/2)f \leq e$, which we have just proved, and multiply both sides by $2/3$ to get $f \leq (2/3)e$. Add $v - e$ to both sides and the result is $v + f - e \leq v + (2/3)e - e$. *G* satisfies the hypothesis of Euler's Formula, so $v + f - e = 2$. Therefore

$$2 \leq v + (2/3)e - e$$
$$2 \leq v - (1/3)e$$
$$(1/3)e + 2 \leq v$$
$$(1/3)e \leq v - 2$$
$$e \leq 3v - 6.$$

In this theorem the hypothesis that $v \geq 3$ is unavoidable. There are only two planar and connected graphs with $v < 3$: K_1 (for which $v = 1$, $f = 1$, and $e = 0$) and K_2 (for which $v = 2$, $f = 1$, and $e = 1$), and the theorem is false for both of them.

Corollary 11. *K_5 is nonplanar.*

Proof. This proof is independent of the one we constructed in chapter 3, and is a lot shorter. Suppose for the sake of argument that K_5 were planar. We know that K_5 is connected so by Theorem 11 K_5 would have the property $e \leq 3v - 6$. But K_5 has $e = 10$ and $3v - 6 = 9$, a contradiction, so K_5 is nonplanar.

Comparing the proof of this corollary to our original proof that K_5 is nonplanar (Theorem 4), we get our first glimmer of the tremendous power of Euler's Formula.

Theorem 12. *If G is planar and connected with $v \geq 3$ and G is not a supergraph of K_3, then $2f \leq e \leq 2v - 4$.*

Proof. Case 1. G has a face bounded by fewer than four edges.
Then—think about this—G must be either P_3 or P_4, or T_4 shown in Figure 101. The theorem holds for all three graphs.
Case 2. Every face of G is bounded by four or more edges.
Then we can make the series of statements

$4 \leq$ the number of edges bounding face 1
$4 \leq$ the number of edges bounding face 2
$4 \leq$ the number of edges bounding face 3
$\vdots$
$4 \leq$ the number of edges bounding face f,

and can conclude as in the proof of the last theorem that $4f \leq 2e$, which upon division by 2 gives the first half of the theorem. Then

$$f \leq (1/2)e$$
$$v + f - e \leq v + (1/2)e - e;$$

Euler's Formula gives $v + f - e = 2$, so

$$2 \leq v + (1/2)e - e$$
$$2 \leq v - (1/2)e$$
$$(1/2)e + 2 \leq v$$
$$(1/2)e \leq v - 2$$
$$e \leq 2v - 4.$$

which is the second half.

As before the hypothesis that $v \geq 3$ is necessary, for the theorem is false for K_1 and K_2.

Corollary 12. *UG is nonplanar.*

Proof. This is similar to the proof of the last corollary. Suppose for the sake of argument that *UG* were planar. *UG* is connected and is not a supergraph of K_3, so by Theorem 12 it would have to be true that $e \leq 2v - 4$. But $e = 9$ and $2v - 4 = 8$, a contradiction, hence *UG* is nonplanar.

In the next section we'll reflect on what has happened in these last two corollaries, but first two more results.

Theorem 13. *If G is planar and connected then G has a vertex of degree ≤ 5.*

Proof. Let *G* be planar and connected.

Case 1. *G* has fewer than three vertices.
  Then *G* must be K_1 or K_2 and the theorem holds.
Case 2. *G* has three or more vertices.
  Suppose for the sake of argument that every vertex of *G* has degree ≥ 6. Then, numbering the vertices of *G* from 1 to v, we can make the series of statements

$$6 \leq \text{the degree of vertex 1}$$
$$6 \leq \text{the degree of vertex 2}$$
$$6 \leq \text{the degree of vertex 3}$$
$$\vdots$$
$$6 \leq \text{the degree of vertex } v.$$

Hence $6v$, the sum of the first column, is less than or equal to $2e$, which is the sum of the second column—see Exercise 11 of Chapter 2. Dividing both sides of this inequality by 2 gives us $3v \leq e$. But by Theorem 11, $e \leq 3v - 6$, so $3v \leq 3v - 6$, a contradiction. Therefore our supposition must be false and *G* must have at least one vertex with degree ≤ 5.

Corollary 13. *If G is planar then G has a vertex of degree* ≤ 5.

Proof. If G is planar and connected it has a vertex of degree ≤ 5 by the theorem; so let G be planar and disconnected.

Disconnected graphs are in two or more connected pieces, as we have seen. Let J be any such piece. Then J is in its own right a planar (by Theorem 5) connected graph, and so by Theorem 13 J has a vertex of degree ≤ 5. This vertex is also a vertex of G and so G has a vertex of degree ≤ 5.

Algebraic topology

In the last section there surfaced two new proofs of the nonplanarity of UG and K_5. The purpose of this section is to shed some light on the nature of these new proofs.

Euclidean geometry is concerned with the "metric" properties of figures, that is those properties that can be measured: lengths of lines, sizes of angles, areas, etc. From a metric perspective Figure 102 contains five different objects.

Topology (not to be confused with "topography") is a sort of generalized geometry. It is a relatively new branch of mathematics wherein the first four objects in Figure 102 are considered to be "the same". Each of these four has a property shared by the other three, but not by the fifth. To see what the property is, imagine that the first object is made of extremely pliable rubber. The rules are that we may stretch, shrink, or otherwise distort the rubber as much as we want, but we may not tear it or fold it back onto itself. Under these rules we could transform the first triangle successively into the right triangle, the square, and the circle, but not into the fifth object. A transformation of one figure into another by this process is called a "continuous deformation". A topologist considers figures that can be continuously deformed into one another to be equivalent; or in other words, a topologist studies only those properties of figures that are preserved by continuous deformations. Being a "continuous

a) b) c) d) e)

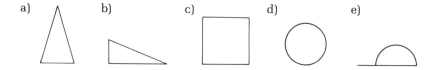

Figure 102

simple closed curve" is an example of such a property (see page 66), and thus the Jordan Curve Theorem is an example of a topological theorem. It says that when any object topologically equivalent to a circle is embedded in a plane, it cuts the plane into two pieces.

Henri Poincaré, one of the founders of the modern phase of topology, described the subject as follows (translation by Edna Kramer, quoted by her in *The Nature and Growth of Modern Mathematics*, vol. 2, p. 326):

> "[Topology] is a purely *qualitative* subject where quantity is banned. In it two figures are always equivalent if it is possible to pass from one to the other by a continuous deformation, whose mathematical law can be of any sort whatsoever as long as continuity is respected. Thus a circle is equivalent to any sort of closed curve but not to a straight-line segment, because the latter is not closed. Suppose that a model is copied by a clumsy craftsman so that the result is not a duplicate but a distortion. Then the model and the copy are not equivalent from the point of view of metric geometry. . . . The two figures are, however, topologically equivalent.
>
> It has been said that geometry is the art of applying good reasoning to bad diagrams. This is not a joke but a truth worthy of serious thought. What do we mean by a poorly drawn figure? It is one where proportions are changed slightly or even markedly, where straight lines become zigzag, circles acquire incredible humps. But none of this matters. An inept artist, however, must *not* represent a closed curve as if it were open, three concurrent lines as if they intersected in pairs, nor must he draw an unbroken surface when the original contains holes."

There's an old anecdote on topology that runs something like this. A topologist enters a coffee shop, orders coffee and a doughnut, and is served. Preoccupied with topological theorems, he takes a bite out of his coffee cup and has to finish his thoughts in a nearby emergency ward. His mistake is somewhat understandable as a doughnut and coffee cup are topologically equivalent, as sketched in Figure 103.

Algebra is another branch of mathematics; it studies sets on which there have been defined things called "operations". An operation on

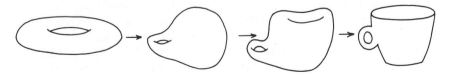

Figure 103

a set is a rule whereby two or more elements of the set can be combined to form another element of the set. High school algebra is the algebra of one specific set, the set of real numbers, and four specific operations defined on that set, addition, subtraction, multiplication, and division. High school algebra is only the tip of the algebraic iceberg.

It so happens that topological problems are somewhat harder to solve than algebraic ones. There is a further factor that algebra has been studied for far longer than topology, so there has been more time to develop algebraic techniques. And so, to make progress in topology while at the same time taking advantage of the advanced state of algebraic knowledge, mathematicians have devised a branch of topology known as *algebraic topology,* wherein algebraic methods are applied to topological problems. In algebraic topology the procedure runs something like this: take a topological problem and if possible convert it into an algebraic problem; try to solve the algebraic problem; if you succeed, reconvert the algebraic solution into topological terms, and the result will be a topological solution to the original problem.

This technique of convert-solve-reconvert is not unique to algebraic topology. It is in fact one of the oldest methods in mathematics. For example, if asked to express the product of CLIII and XXIX as a Roman numeral, you would probably *convert* the original problem CLIII × XXIX into the corresponding Hindu-Arabic numeral problem 153 × 29, *solve* the Hindu-Arabic problem in the standard way, and then *reconvert* the Hindu-Arabic solution 4437 into the Roman numeral MMMMCCCCXXXVII which is the solution to the original problem. Analytic geometry, developed in 1637 by René Descartes, is another example of the convert-solve-reconvert technique. In analytic geometry we take a problem in geometry, convert it into a problem about a system of equations, solve the system of equations, and then reconvert the "analytic" (actually algebraic) solution back into geometric terms.

In order to use the convert-solve-reconvert technique there must be a *means* whereby problems in one area of mathematics can be converted into problems in another area. In the case of the Roman numeral problem the means was the familiar one learned in grammar school. In the case of analytic geometry the means is to associate with every geometric point a pair (x,y) of real numbers called the "coordinates" of the point; straight lines are thereby converted into equations, triangles into systems of equations, and so on.

In the last section when we proved anew that UG and K_5 are nonplanar, we were doing a bit of algebraic topology. Graph theory is a small part of topology, high school algebra is a small part of algebra, and Euler's Formula is the means whereby we converted problems in graph theory to problems in high school algebra. In Corollary 11 for example, the statement is purely topological, yet the proof, which is based on Euler's Formula via Theorem 11, is purely algebraic. Similar remarks hold for Corollary 12. In these situations our work was algebraic but its import was topological.

In the sequel we will be doing more algebraic topology, so it might be well at this point to mention that subject's chief advantage and chief disadvantage. The advantage is that an algebraic proof of a topological theorem is shorter than a topological proof, which may not even exist. The disadvantage is that an algebraic proof is less conducive to real understanding.

Take for example our two proofs that UG is nonplanar. The first proof (Theorem 3) was purely topological. It was quite long, and even at that was essentially incomplete, for we accepted the Jordan Curve Theorem without proof. The second proof (Corollary 12) was purely algebraic, quite short, and complete. Yet the second proof is somehow mysterious. After reading it we wonder what happened. Our intellect is convinced but our intuition isn't. We leave the second proof with no real *feeling* as to why UG is nonplanar. The proof is so concise that all intuitive handles have been squeezed out. Deep down our intuition wonders if it hasn't been tricked. There is none of this mystery in the first proof. The Jordan Curve Theorem is so intuitively "obvious" that its rigorous justification is not missed. Because of the length of the proof, and the diagrams, and the fact that each step is basic and not dependent on an elaborate substructure of previous theorems, our intuition is satisfied along with our intellect.

Most people would judge that the advantage outweighs the disadvantage. And for people who like mystery stories, or who fantasize being seduced by a mysterious stranger, what I have called the "disadvantage" isn't a disadvantage at all.

Exercises

1. N_1 is certainly planar, and we proved on page 98 that it is connected. Prove now that it is polygonal by proving that the statement "every edge of N_1 borders on two different faces" is true.
2. Find the error in the following "proof":

   Theorem. *If A is a set of horses, then all the horses in A are of the same color.*

   Proof. Let S be the statement, "if A is a set of n horses, then all the horses in A are of the same color." S is a statement about positive integers n, and we shall prove that S is true for every positive integer by the principle of mathematical induction; this will establish the theorem. The first step is to prove

(1) S is true for $n = 1$.

Rephrased, this is "if A is a set containing one horse, then all the horses in A are of the same color." Obviously this is true. Next we have to prove

(2)   if S is true for $n = k$ then S is true for $n = k + 1$.

That is, we are given the truth of "if A is a set of k horses, then all the horses in A are of the same color," and we have to deduce from this the truth of "if A is a set of $k + 1$ horses, then all the horses in A are of the same color." So let A be a set of $k + 1$ horses. For the sake of reference we shall number the horses from 1 to $k + 1$.

Remove horse #1. There remains a set of k horses. We are given that all the horses in any set of k horses are of the same color, so horse #2, horse #3, ..., and horse #$(k + 1)$ must be all of the same color. Now replace horse #1 and remove horse #$(k + 1)$. Again we are left with a set of k horses, so horse #1, horse #2, ..., and horse #k must be all of the same color. Obviously then all the horses in A, from #1 to #$(k + 1)$, must be of the same color.

Invoking the principle of mathematical induction it follows that S is true for every positive integer and the theorem is proved.

Corollary. *All the horses in the world are of the same color.*

Proof. Let A be the set of all the world's horses, and apply the theorem to A.

This corollary can be used to prove a wide range of things, for example

Theorem. $1 + 1 = 3$.

Proof. If $1 + 1 = 3$, fine. If not, that would be a horse of a different color, which doesn't exist by the corollary.

3. Believe it or not, the graph of Figure 104a is planar. Find its number of faces. (If you use Euler's Formula you won't need to draw it without edge-crossings.)

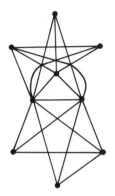

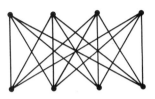

Figure 104

4. Imitate the proof of Corollary 12 to construct a proof, independent of the results of Chapter 3, that the graph of Figure 104b is nonplanar.

5. Crucial to the proof of Euler's Formula, the following step involves more than is immediately apparent:

> Now let G be an arbitrary polygonal graph having $k + 1$ faces. Remove some of the edges and vertices bordering the infinite face of G to produce a new polygonal graph H having one less face than G, so H has k faces.
> —page 103

Find a polygonal graph G having a face bordering the infinite face which, if removed, results in a subgraph H which is not polygonal. This shows that we must be choosy about the exterior face of G that is removed.

6. Prove this partial converse of Euler's Formula: if a graph is planar and $v + f - e = 2$, then the graph is connected.

7. **Definition.** A *component* of a graph is a connected subgraph that is not contained in a larger connected subgraph.

Thus a connected graph is its own single component, and the components of a disconnected graph are what we have been calling the "pieces" that comprise it.

Examples. Figure 105a has three components, which are displayed separately in 105b, c), and d). Figure 106 has only one component, itself.

Let "p" denote the number of components of a graph, and prove this generalization of Euler's Formula: if a graph is planar, then $v + f - e = 1 + p$.

a) 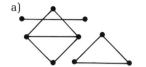　　　b) 　　c) 　　d)

Figure 105

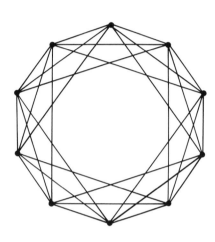

Figure 106

8. Corollary 13 can sometimes be used to prove a graph nonplanar. Use it on the graph of Figure 106.

9. Show by example that the statement "every planar graph has a vertex with degree less than or equal to 4" is false. That is, find a planar graph in which every vertex has degree greater than or equal to 5. This shows that, in this fashion at least, Corollary 13 can't be improved upon, that it is already the strongest statement that can possibly be made.

10. There is another manner, however, in which Corollary 13 can be improved upon. Prove: Every planar graph with $v \geq 4$ has at least *four* vertices of degree ≤ 5.

11. **Definition.** The *connectivity* of a graph is the smallest number of vertices whose removal (together with their incident edges) results in either K_1 or a disconnected graph. We shall denote the connectivity of a graph by "c".

Examples. Removing vertices with their incident edges never disconnects K_4, but after removing three vertices (with edges) we are left with K_1, so K_4 has $c = 3$. The graph of Figure 105a has $c = 0$ as it is already disconnected. Figure 105b has $c = 2$, 105c has $c = 1$, 105d has $c = 2$, and 106 has $c = 6$.

The connectivity of a graph indicates the *extent* to which it is connected, in some sense. Note that c is a *minimum*. If you were to remove successively vertices 3, 1, 6, and 4 from Figure 79b you might erroneously conclude that the graph has $c = 4$, as the debris was disconnected only after the fourth removal. But starting over and removing vertices 2 and 9 shows that c can be at most 2. c is exactly 2 since the removal of any one vertex leaves a connected subgraph. Find c for each of the graphs in Figures 91, 92, and 93.

12. By Exercise 11 of Chapter 2, $2e/v$ is the average of the degrees of a graph. Prove that if a graph has connectivity c then c is less than or equal to $2e/v$.

13. Use the previous exercise to show that there is no graph with $e = 7$ and $c = 3$, and none with $e = 11$ and $c = 4$.

14. **Definition.** A *bridge* in a graph is an edge whose removal would increase the number of components.

Example. In Figure 107 there are six bridges: $\{1,2\}$, $\{12,13\}$, $\{4,5\}$, $\{6,7\}$, $\{7,8\}$, and $\{18,19\}$. The graph has two components; if any bridge were removed the resulting subgraph would have three components.

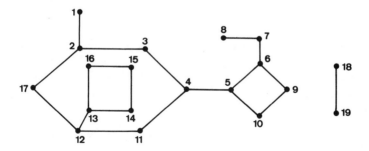

Figure 107

In a planar graph a bridge necessarily borders on only one face, and an edge bordering on only one face is necessarily a bridge. Thus bridges are the things that prevent planar connected graphs from being polygonal. Use this fact to prove that if a planar and connected graph G has the property that the boundary of every face is a cyclic graph, then G is polygonal. Then show that the converse statement is false by finding a polygonal graph having a face whose boundary is not a cyclic graph.

15. By Theorem 11 we know that every planar, connected graph with $v \geq 3$ has $e \leq 3v - 6$. Prove that if such a graph G has the additional property that every supergraph of G with one more edge is nonplanar, then the boundary of every face of G is C_3, G is polygonal, and G has $e = 3v - 6$.

16. Prove: if G is planar and connected with $v \geq 3$ and the boundary of every face is C_4, then G is polygonal and $e = 2v - 4$.

17. Prove: if the connectivity c of a graph is at least 6, then the graph is nonplanar.

18. Prove: if a nonplanar graph has $v \geq 6$, $c \geq 3$, and a subgraph which is an expansion of K_5, then it also has a subgraph which is an expansion of UG. This is what I had in mind on page 86 when I advised you that "the vast majority of nonplanar graphs contain an expansion of UG, so start by looking for that."

Suggested reading

On Leonhard Euler

Men of Mathematics by Eric Temple Bell (Simon and Schuster, paperback edition 1962), chapter 9.

On Topology

What is Mathematics? by Richard Courant and Herbert Robbins (Oxford University Press, 1941), chapter V. One of the best all-around mathematics textbooks since Euclid. Chapter V has been excerpted in volume 1 of *The World of Mathematics*, edited by James R. Newman (Simon and Schuster, 1956).

Intuitive Concepts in Elementary Topology by B. H. Arnold (Prentice-Hall, 1962).

Topology by E. M. Patterson (Interscience Publishers, 1959).

*"Topology" by Albert W. Tucker and Herbert S. Bailey, Jr. in the January, 1950 issue of *Scientific American*; reprinted as chapter 19 of *Mathematics in the Modern World: Readings from Scientific American* with introductions by Morris Kline (W. H. Freeman, 1968).

5. PLATONIC GRAPHS

Introduction

Platonic graphs are fun to talk about for three reasons. The first is historical: the five most interesting platonic graphs are identifiable with the five so-called "platonic solids" (after Plato) of ancient mathematics and mysticism. The second is heuristic: the theory of platonic graphs is a spectacular warning to mathematicians of what can happen if they overindulge their natural tendency to place conditions on the objects they are studying. And the third is pedagogical: the theorem proved in this chapter is a striking demonstration of the power of Euler's Formula.

Definition 25. A graph is *regular* if all the vertices have the same degree. If the common value of the degrees of a regular graph is the number d, we say that the graph is *regular of degree d*.

Examples. 1) Any cyclic graph C_v is regular of degree 2.
2) Any complete graph K_v is regular of degree $v - 1$.
3) Any null graph N_v is regular of degree 0.
4) UG is regular of degree 3.
5) Figure 108a is regular of degree 3.
6) Figure 108b is regular of degree 4.

Figure 108

Definition 26. A graph is *platonic* if it is polygonal, regular, and has the additional property that all of its faces are bounded by the same number of edges.

Examples. Let us search among the previous examples for platonic graphs.

1) Each C_v is platonic. It is polygonal and regular, and each of its two faces is bounded by v edges.

2) Of the complete graphs only K_1, K_3, and K_4 are polygonal. All three are regular. K_1 has only one face, that bounded by 0 edges, so K_1 is platonic, though shallowly. K_3 is the same as C_3 and is platonic. Each of the four faces of K_4 is bounded by 3 edges, so K_4 is platonic as well.

3) N_1 is the only polygonal null graph. It is the same as K_1 and so is platonic.

4) *UG* is not polygonal, so it is not platonic.

5) Figure 108a is not polygonal because the central edge borders only the infinite face. Even if it were polygonal it would still not be platonic because it has faces bounded by 3, 4, and 9 edges.

6) Figure 108b is a supergraph of *UG*. Hence it is nonplanar, hence it is not polygonal, hence it is not platonic.

There has been a steady escalation of conditions since our first discussions in Chapter 2. Then we talked about plain old graphs. Subsequently we restricted our attention to planar graphs, then to planar connected graphs, then to planar connected graphs with each edge bordering two faces (polygonal graphs), and now to planar connected regular graphs with each edge bordering two faces and all faces bounded by the same number of edges (platonic graphs). Whew! Every time a new condition is added we find ourselves talking about fewer things than before. This process of piling condition on condition can lead to ridiculous results. The term "platonic" is a case in point, as there are very few platonic graphs. In fact, other than the ones we have already discovered—K_1, K_4, and the cyclic graphs C_v—there are only four! We now turn our attention to proving this remarkable fact.

Proof of the theorem

Lemma 14. *If G is regular of degree d then* $e = dv/2$.

Proof. Left to the reader. A "lemma" is an auxiliary theorem used to prove another theorem.

Lemma 15. *If G is a platonic graph, d is the degree of each vertex, and n is the number of edges bounding each face, then* $f = dv/n$.

Proof. Left to the reader.

Theorem 16. *Other than* K_1 *and the cyclic graphs there are only five platonic graphs.*

Proof. Let G be a platonic graph which is not K_1 or a cyclic graph. Let d be the degree of each vertex of G and let n be the number of edges bounding each face.

If d were 0, G would be a platonic null graph, i.e., K_1. But G is not K_1 so $d \neq 0$.

If d were 1, then G would not be polygonal—see Exercise 1. But G is polygonal so $d \neq 1$.

If d were 2, then G would be a cyclic graph—see Exercise 2. But G is not cyclic so $d \neq 2$.

Thus we conclude that $d \geq 3$. Note that n, being the number of edges bounding a face of a polygonal graph, is also at least 3.

The proof is now half over. The remaining half consists in substituting the formulae of the two lemmas into Euler's Formula, performing a few algebraic manipulations, and interpreting the result.

G is planar and connected so $v + f - e = 2$. By the lemmas, $e = dv/2$ and $f = dv/n$. Thus

$$v + dv/n - dv/2 = 2.$$

Multiplying both sides by $2n$ we get

$$2nv + 2dv - ndv = 4n$$
(1)
$$v(2n + 2d - nd) = 4n.$$

v and $4n$ are both positive numbers so $(2n + 2d - nd)$ must be positive also; that is,

$$2n + 2d - nd > 0$$
$$-(2n + 2d - nd) < 0$$
$$nd - 2n - 2d < 0$$
$$nd - 2n - 2d + 4 < 4$$
$$(n - 2)(d - 2) < 4.$$

The last inequality is the punch line. It was deduced from the premise that we were dealing with a platonic graph other than K_1 or a cyclic graph, hence the "d" and "n" of *every* platonic graph, other than K_1 or a cyclic graph, must satisfy $(n - 2)(d - 2) < 4$. But $(n - 2)(d - 2) < 4$ has only five solutions for which $d \geq 3$ and $n \geq 3$, and to each of these there corresponds but one platonic graph.

The table on p. 121 was constructed as follows. The first two columns contain all possible combinations of d and n that satisfy $(n - 2)(d - 2) < 4$ together with our earlier discoveries that $d \geq 3$ and $n \geq 3$. A little reflection will show that there are no more possibilities. Think about this until you are convinced.

Then, knowing particular values for d and n, v can be computed from equation (1), which can be rewritten

$$v = \frac{4n}{2n + 2d - nd}.$$

Thus the third column. Lemma 14 tells you how to compute e from d and v, and this was done to produce the fourth column. Finally, the values of f in the fifth column were computed from the values of d, v, and n using Lemma 15. To each row of the table there corresponds one and only one platonic graph, whose traditional name is given in the last column. These graphs have been drawn in Figure 109; we will explain their names in the next section.

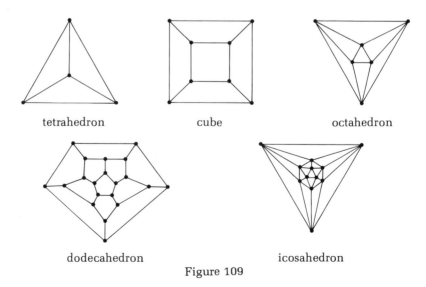

tetrahedron   cube   octahedron

dodecahedron   icosahedron

Figure 109

d	n	v	e	f	name
3	3	4	6	4	tetrahedron or K_4
3	4	8	12	6	hexahedron or cube
3	5	20	30	12	dodecahedron
4	3	6	12	8	octahedron
5	3	12	30	20	icosahedron

It is apparent from Figure 109 that to each row of the table there corresponds at least one platonic graph. Though not apparent it is also true that to each row of the table there corresponds *only* the one platonic graph named in the last column. This requires proof.

In principle we could construct such a proof as follows. For each row of the table, we could look at the numbers in the "*v*" and "*e*" columns, draw all graphs having that number of vertices and that number of edges, and check that the one named in the last column is the only one that is platonic. In practice such a proof would be reasonable for only the first and fourth rows of the table (see Exercise 3). The next simplest row is the second, and there are 1,312 graphs with eight vertices and twelve edges! As you might expect therefore, there are shorter ways of proving that the graphs of Figure 109 are the only platonic graphs corresponding to the table, but we shall pass them over.

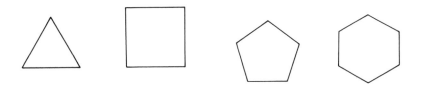

Figure 110

History

In the terminology of geometry a "regular polygon" is a polygon—a closed plane figure bounded by a finite number of straight lines—having all its sides the same length and all its angles the same size. **There are infinitely many regular polygons, the first four of which are drawn in Figure 110.** Considered as a graph, a regular polygon with *v* sides is isomorphic to the cyclic graph C_v.

Still in the terminology of geometry, a *regular polyhedron* is a polyhedron—a closed solid figure bounded by a finite number of planes—having congruent regular polygons for its faces and all of its corner angles the same size. A cube is a regular polyhedron.

The notion "regular polyhedron" is an extension to three dimensions of the notion "regular polygon". In cases like this mathematical experience suggests that since there are infinitely many regular polygons, there are probably infinitely many regular polyhedra as well.

The Pythagoreans (about 500 B.C.) seem to have been the first to discover, if not to prove, the surprising fact that contrary to expectations, there are only five regular polyhedra: the regular tetrahedron, the regular hexahedron (or "cube"), the regular octahedron, the regular dodecahedron, and the regular icosahedron. ("Hedron" means "face"; the prefixes mean respectively "four", "six", "eight", "twelve", and "twenty".)

The five regular polyhedra have been drawn in Figure 111. Considered as three-dimensional drawings of graphs, they are isomorphic to the graphs of Figure 109. Thus, except for K_1, which is platonic in only a trivial sense, every platonic graph corresponds to either a regular polygon or a regular polyhedron.

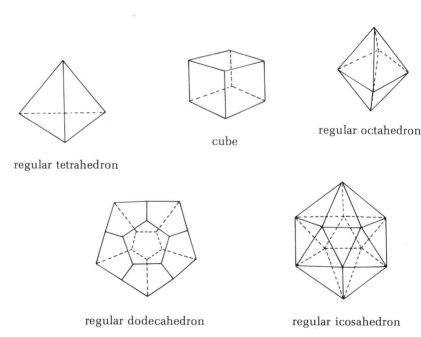

regular octahedron

cube

regular tetrahedron

regular dodecahedron   regular icosahedron

Figure 111

The Pythagoreans were mystics, and believed that the tetrahedron, cube, octahedron, and icosahedron respectively underlay the structure of the four elements of Greek science: fire, earth, air, and water. The dodecahedron they identified with the universe as a whole. Plato was quite taken by all this and spent some time in his dialogue *Timaeus* (named after the Pythagorean who is the chief interlocutor) discussing the connection between the five regular polyhedra and the structure of the universe. For this reason the regular polyhedra came to be known as the "platonic solids".

The final book of Euclid's *Elements* is devoted to the five regular polyhedra. Some scholars argue that this analysis was Euclid's goal in composing the *Elements.*

Johannes Kepler (1571–1630), a German astronomer-astrologer contemporary with Shakespeare, discoverer of three laws of planetary motion that still bear his name, was seduced at an early age by Pythagorean mysticism. So strong was his faith in the mathematical harmony of the universe that he quickly became convinced of a relationship between the five regular polyhedra and the six known planets. At that time planetary orbits were thought to be circular, and Kepler imagined the circles as equators of huge spheres. He found that if a cube were inscribed in the sphere of Saturn, and another sphere inscribed in the cube, the second sphere would be the sphere of Jupiter; that if a regular tetrahedron were inscribed in the sphere of Jupiter, and a sphere inscribed in the tetrahedron, this third sphere would be the sphere of Mars; and so on down the line, with the five regular polyhedra nesting perfectly among the spheres of the six planets. Kepler concluded that the five regular polyhedra accounted for both the number of planets—there could be only six—and their spacing in the solar system. When his own later work revealed that (1) the spacing of the planetary orbits did not correspond to the nesting of the polyhedra as accurately as he had thought, and (2) planetary orbits are not circular but elliptical, Kepler was forced to put aside his beliefs about the regular polyhedra, but only after great mental anguish.

Exercises

1. Draw all connected graphs that are regular of degree 1.
2. Satisfy yourself that every connected graph which is regular of degree 2 is a cyclic graph. Show by example that deleting the word "connected" results in a false statement.

3. It's easy to see that K_4 is the only platonic graph corresponding to the first row of the table on page 121, as K_4 is the only graph having $v = 4$ and $e = 6$ (see Figure 45). Now verify that there is only one platonic graph corresponding to the fourth row by drawing all graphs with $v = 6$ and $e = 12$ (there are five of them) and checking that only the octahedron is platonic.

4. Prove: there is no regular graph with $v = 6$ and $e = 10$. (This can be done algebraically, without drawing a single graph.)

5. Prove: if a graph has an odd number of vertices and is regular of degree d, then d must be even.

6. Find a graph other than the cube that has $v = 8$ and is regular of degree 3.

7. Draw all regular graphs with six vertices or less. There are twenty of them.

8. **Definition.** A *dual graph* of a planar graph is formed by taking a crossing-free plane drawing of the planar graph, placing a dot inside each face, and joining two dots whenever the borders of the corresponding faces have one or more edges in common.

Example. The planar graph of Figure 112a has been redrawn in 112b along with a dual; the dual is drawn by itself in 112c.

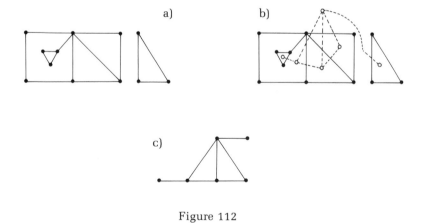

Figure 112

Note that only a planar graph can have a dual, and that a dual is always planar and connected. But a dual is not always unique: some planar graphs have several different duals, each arising from a different crossing-free plane drawing. This is why in the definition I said "a" dual instead of "the" dual.

Example. Figure 113a is isomorphic to Figure 112a, yet the dual formed in 113b and shown by itself in 113c is not isomorphic to the dual of Figure 112c.

Each of the platonic graphs does, however, have a unique dual. For example, K_1 has only one dual, itself, and each cyclic graph has only one dual, K_2. The octahedron and its unique dual are shown in the three-dimensional drawing of Figure 114. Draw the duals of the other four platonic graphs. Your results will help explain the curious symmetry of the table on page 121.

9. Prove that every wheel graph W_v (see Chapter 2, Exercise 6) is isomorphic to its (unique) dual.

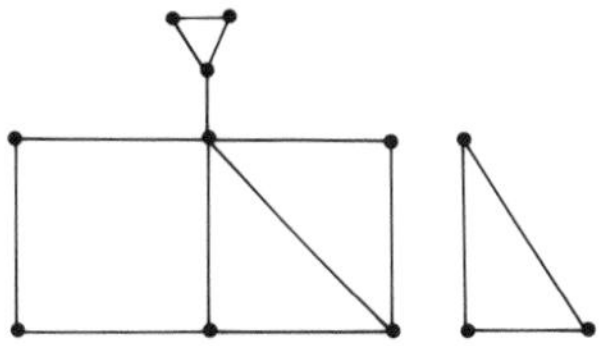

(a)

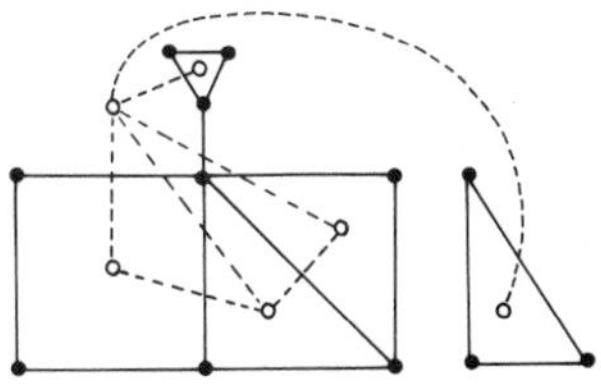

(b)

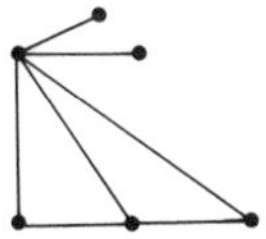

(c)

Figure 113

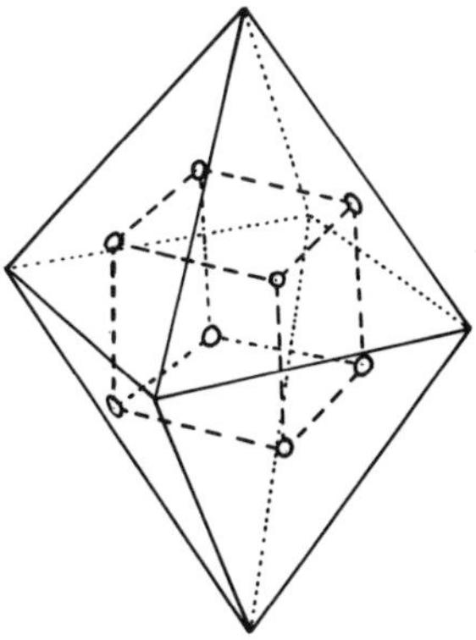

Figure 114.

Suggested reading

On the Regular Polyhedra

*"The Five Platonic Solids" by Martin Gardner, reprinted as Chapter 1 of *The Second Scientific American Book of Mathematical Puzzles and Diversions* by Martin Gardner (Simon and Schuster, 1961).

100 Great Problems of Elementary Mathematics, Their History and Solution by Heinrich Dorrie (Dover, 1965), Problem 71. Dorrie proves that there are only five regular polyhedra using spherical geometry.

On Johannes Kepler

* *The Watershed* by Arthur Koestler (Doubleday Anchor, 1960). Chapter 2, "The Perfect Solids," is an account of Kepler's attempt to apply the regular polyhedra to the problem of planetary orbits.

Kepler by Max Caspar (Collier Books, 1962).

6. COLORING

Chromatic number

Having seen what happens when we pile a great many conditions on the graphs under discussion, we will now abandon the escalation of conditions and return to plain old graphs. Despite the title, crayons will not be required for this chapter.

Definition 27. A graph has been *colored* if a color has been assigned to each vertex in such a way that adjacent vertices have different colors.

In other words, a graph has been colored if each edge has two differently colored endpoints.

Examples. 1) Figure 115 shows two colorings of the cube.

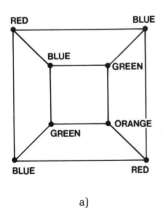
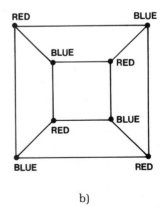

a)   b)

Figure 115

2) More often colors are denoted by numbers. An unlabeled graph that has been colored looks a lot like a labeled graph, but the context should prevent any confusion. The graphs in Figure 116 have been colored; to verify this all you have to do is notice that vertices with the same color are never adjacent.

Definition 28. The *chromatic number* of a graph is the smallest number of colors with which it can be colored. We shall denote the chromatic number of a graph by "X".

It must seem strange to use the letter "X", but we've already used "c" (Exercise 11, Chapter 4) and I think of "X" as being the Greek letter chi (χ).

Examples. 1) The cube has $X = 2$. Obviously it requires at least two colors, and Figure 115b shows that two are enough.

2) Graph a) in Figure 116 has $X = 3$. It needs at least three colors

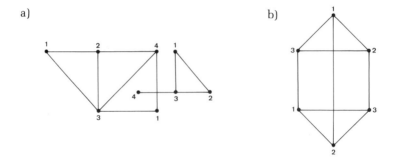

Figure 116

because it is a supergraph of K_3 and K_3 needs three. That three colors are sufficient to color graph a) can be seen from Figure 117.

3) Similarly graph b) of Figure 116 needs at least three colors because it is a supergraph of K_3. As it is obvious from the figure that three colors suffice, graph b) has $X = 3$.

4) It is natural to assume that nonplanar graphs have bigger chromatic numbers than planar graphs. In the broad sense, this is true; that is, if a nonplanar graph and a planar graph are chosen at random, it is likely that the nonplanar graph will have the larger chromatic number. However, there are infinitely many exceptions. For example, the thoroughly nonplanar graph of Figure 118a has $X = 2$, while the simple planar graph of 118b has $X = 4$.

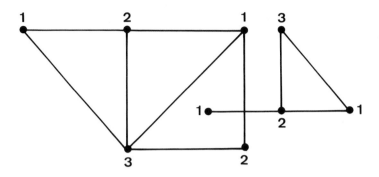

Figure 117

a)

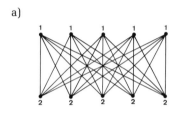

b)

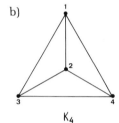

K_4

Figure 118

A graph can always be colored by giving a different color to every vertex, so for any graph $X \le v$. And of course X is always at least 1, so the chromatic number of a graph must satisfy the inequality $1 \le X \le v$. The complete graphs are the only graphs for which X is actually equal to v, as any other graph has at least one pair of nonadjacent vertices, which can be given the same color; and the null graphs are the only graphs with $X = 1$, as any other graph has at least one edge, which must have its endpoints differently colored. With these exceptions therefore, the chromatic number of a graph must satisfy the inequality $2 \le X \le v - 1$.

For an arbitrary graph there is no simple rule for determining where within the range $2 \le X \le v - 1$ the chromatic number falls. In most cases the thing to do is color the graph as economically as you can and then study the result until you are certain that no coloring is possible with fewer colors. For a complicated graph it may be necessary to write down a proof that it cannot be colored with fewer colors.

Coloring planar graphs

It is a simple matter to find planar graphs with $X = 1$, 2, 3, or 4. For example N_7, the cube, graph b) of Figure 116, and K_4 are planar graphs that have respectively those chromatic numbers. But no one has ever found a planar graph with X larger than 4; so it appears that planar graphs always have $X \leq 4$. No one has succeeded in proving this, though the prevailing mathematical opinion is that it is true. A statement that mathematicians can prove is called a "theorem", or sometimes a "lemma" or "corollary" if it bears a certain relationship to another theorem. A statement that mathematicians believe but cannot as yet prove is called a "conjecture". A conjecture is an unsolved problem. The task is to either prove that the conjecture is true, at which point the conjecture is promoted to "theorem", or show by example that the conjecture is false. If either thing happens the conjecture has been "resolved". A few sentences ago we made a conjecture about the chromatic number of a planar graph. Let us now state this conjecture formally, with its traditional name.

The Four Color Conjecture. *Every planar graph has $X \leq 4$.*

Because of its accessibility—the conjecture can be explained to anyone in a couple of minutes—and its persistent failure to be resolved—mathematicians have been trying for more than 120 years—the Four Color Conjecture is one of the most famous unsolved problems in mathematics.

In 1879 a mathematician named A. B. Kempe published a purported proof of the Four Color Conjecture, only to have it shot down ten years later by P. J. Heawood. But Heawood's criticism of Kempe's "proof" was not solely destructive, for he showed how the previous attempt could be modified into a satisfactory proof of what is now called the Five Color Theorem.

The Five Color Theorem. *Every planar graph has $X \leq 5$.*

This is a theorem, not a conjecture. It is provable; in fact, we shall prove it shortly. The Five Color Theorem guarantees that there are no planar graphs with $X > 5$. But we are still left with the question, "Are there any planar graphs with $X = 5$?" No one knows. The Four Color Conjecture stands unscathed as a most maddening puzzle.

Periodically mathematical circles are electrified with the rumor that someone has resolved the Four Color Conjecture. But careful examination of the proposed proofs has always disclosed an error. In a way, I think, it would be a shame if the conjecture were ever resolved,

for the suspense and mystery would evaporate. *

Like few other mathematical problems, the Four Color Conjecture offers the average nonmathematician a chance at a place in the annals of mathematical achievement. If the conjecture is true, of course, a proof of that fact is probably beyond most people, including most mathematicians; and admittedly the prevailing opinion is that the Four Color Conjecture is true. But if it is in fact false—prevailing mathematical opinion has been wrong before—then all that is required to show this is the discovery of a planar graph with $X = 5$. If such a graph exists and is not monstrous—say, if it has fewer than 100 vertices—then it might be discovered by almost anyone. You might enjoy trying to construct such a graph, sometime between now and the end of your life. Once again, the task is simply to draw a planar graph that requires five colors. Success will bring instant fame. One qualification: in 1969 the Four Color Conjecture was proved for planar graphs with $v \le 39$, so any planar graph requiring five colors must have at least 40 vertices.

Proof of the Five Color Theorem

The proof of the Five Color Theorem is rather long and involved. In order to comprehend it, you may have to peruse it several times. But I think the investment of brain-strain is worthwhile, for your own intellect is far more compelling than outside authority.

The Five Color Theorem. *Every planar graph has* $X \le 5$.

Proof. The proof is by mathematical induction on v. Let S be the statement, "every planar graph with v vertices has $X \le 5$." Then S is a statement about positive integers and we will have the Five Color Theorem if we can prove that S is true for every positive integer. By the principle of mathematical induction it will suffice to prove the statements numbered (1) and (2) below. As in most induction proofs, the first is a snap and the second more difficult.

(1) S is true for $v = 1$.

That is, we have to prove that every planar graph with one vertex has $X \le 5$. The only planar graph, indeed the only graph, with one vertex is N_1, and for this graph $X = 1$.

We might note in passing that S is just as clearly true for $v = 2, 3, 4,$ or 5 since the chromatic number of a graph cannot exceed its number of vertices.

*Alas! See Afterword, p. 195.

(2)   If S is true for $v = k$, then S is true for $v = k + 1$.

We are thus given that every planar graph with k vertices has $X \leq 5$, and the task is to deduce that every planar graph with $k + 1$ vertices has $X \leq 5$. So let G be an arbitrary planar graph having $k + 1$ vertices, and we shall do our best to prove that G has $X \leq 5$.

Corollary 13 (page 108) guarantees that G possesses at least one vertex, call it "A", such that the degree of A is ≤ 5. Denote by "$G - A$" the subgraph of G obtained by erasing vertex A and all edges incident to vertex A. Then $G - A$ is a planar graph having k vertices. We are given that all such graphs have $X \leq 5$. So $G - A$ can be colored with five colors or less. Suppose that this has been done. All that remains is to extend the coloring of $G - A$ with at most five colors to a coloring of G with at most five colors.

Case 1. $G - A$ has $X < 5$.

In this case $G - A$ has been colored with at most four colors so G, which is obtained from $G - A$ by restoring vertex A and its incident edges, can certainly be colored with at most five. Simply color vertex A with a color not used for $G - A$.

Case 2. $G - A$ has $X = 5$ and A has degree < 5.

In this case, vertex A is adjacent to at most four vertices of G. Color vertex A with a color not used for its adjacent vertices, and the result is a coloring of G with five colors.

Case 3. $G - A$ has $X = 5$ and A has degree 5.

Let P, Q, R, S, and T be the five vertices adjacent to A. If any two of these five vertices have been colored with the same color, then there is a color available for A and we have a coloring of G with five colors. So we may as well assume that P, Q, R, S, and T have been colored respectively with five different colors 1, 2, 3, 4, and 5.

This case is the difficult one. Now there is no obvious way of coloring vertex A without resort to a sixth color. It can be done, however, if we can recolor one of the vertices adjacent to A without unduly upsetting the coloring of $G - A$.

Subcase i of case 3. There is no walk joining P to R consisting entirely of vertices colored 1 or 3.

Figure 119 depicts what the relevant portion of G might look like. In the figure it is assumed that all the walks touching either P or R, and consisting entirely of vertices colored 1 or 3, are shown. Note that there is no way of getting from P to R by travelling along such a walk.

Recolor P with color 3, and interchange the colors of the vertices

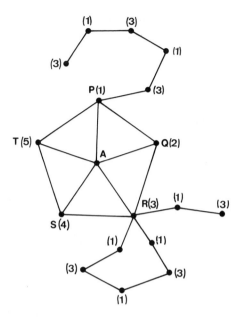

Figure 119. Relevant portion of G in subcase i of case 3

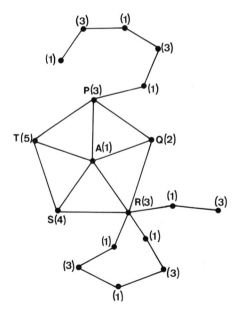

Figure 120. Same portion after color interchange

of all color 1-color 3 walks touching P. This has been done in Figure 120. Because of the assumption under which we are working, R and the vertices of the color 1-color 3 walks touching R are unaffected by this. Now color 1 has been freed for vertex A and we have a coloring of G with five colors.

Subcase ii of case 3. There is a walk joining P to R consisting entirely of vertices colored 1 or 3.

The trick of the last subcase won't work this time, for recoloring P with color 3 will force a recoloring of R with color 1, and vertex A will still be adjacent to five differently colored vertices.

Figure 121 depicts what the relevant portion of G might look like. In the figure it is assumed that all walks touching either Q or S, and consisting entirely of vertices colored 2 or 4, are shown. Note that there is no color 2-color 4 walk joining Q to S. That this is necessarily so can be seen by the following argument.

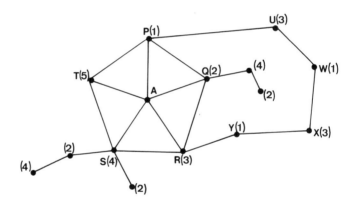

Figure 121. Relevant portion of G in subcase ii of case 3

G is planar, so we may as well assume that our drawings have no edge-crossings. Vertex Q is surrounded by a cyclic subgraph of $G—PUWXYRAP$ in Figure 121—none of whose vertices is colored 2 or 4; vertex A is not colored at all, and the others have been colored 1 or 3. Since G is planar any color 2-color 4 walk joining Q to S must pass through one of the vertices of this cyclic subgraph (by the Jordan Curve Theorem). But no vertex of the cyclic subgraph has been colored 2 or 4, thus there is no walk joining Q to S consisting entirely of vertices colored 2 or 4.

Now we can use the trick of the last subcase to recolor vertex Q

with color 4, interchanging the colors of the color 2-color 4 walks touching Q accordingly. Vertex S will be unaffected by this recoloring, so vertex A can be colored with color 2 and we have a coloring of G with five colors—see Figure 122.

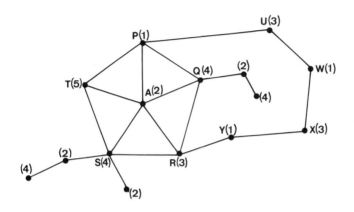

Figure 122. Same portion after color interchange

G was an arbitrary planar graph with $v = k + 1$. In each case we have shown that G can be colored with at most five colors, so G has $X \leq 5$. This completes the proof of statement (2) and we're finished.

The key to the proof of the Five Color Theorem is Euler's Formula, though the length of the proof and the number of intermediary theorems makes this easy to overlook. The substance of the proof is of course the proof of statement (2). Our proof of statement (2) depends entirely on knowing that G has a vertex of degree less than or equal to 5. This is Corollary 13, which we derived from Theorem 13; and Theorem 13 is a derivative of Theorem 11, which is in turn a derivative of Euler's Formula.

It's clear that if the Four Color Conjecture is ever proved, the proof will be substantially different from the proof of the Five Color Theorem and not merely modeled on it. For to try to prove the Four Color Conjecture by imitating the proof of the Five Color Theorem, we would need the following analog of Corollary 13—"every planar graph has at least one vertex with degree less than or equal to 4"; but as we mentioned in Exercise 9 of Chapter 4, that statement is false—for example, the icosahedron is a planar graph, but every vertex has degree 5. Imitating the proof of the Five Color Theorem is not entirely fruitless, however, for by doing so we can prove the following more modest theorem, which tells us where to look for counterexamples to the Four Color Conjecture.

Theorem 17. *Every planar graph having a vertex of degree* ≤ 4 *has* $X \leq 4$.

Proof. Imitate the proof of the Five Color Theorem.

Since the existence of planar graphs with all degrees ≥ 5 is what blocks the most natural avenue to a proof of the Four Color Conjecture, these graphs take on a special significance as test cases. They are the only planar graphs that could possibly have $X = 5$.

Coloring maps

Figure 123a is a map of South America. Letting borders be "edges" and border junctions be "vertices" we can interpret the map as a planar graph, which graph is isomorphic to the graph of Figure 123b. In a similar fashion any other map, actual or imaginary, can be interpreted as a planar graph. Conversely any planar graph can be interpreted as a map, usually an imaginary one, by letting the faces, including the infinite face, be "countries". If your imagination balks at having a country surround several others, think of the infinite face as an "ocean". Under these interpretations maps and planar graphs are the same things.

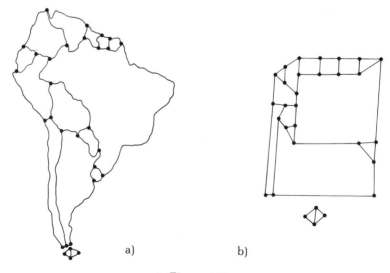

a)   b)

Figure 123

Mapmakers usually color their maps in such a way that countries—including the ocean as a "country"—sharing a border have different colors. Since the cost of printing a map goes up with the number of colors used, mapmakers are naturally interested in knowing the minimum number of colors with which a given map can be colored. Generalizing a bit we are led to consider the following problem.

Map coloring problem. Find the smallest number m such that the faces of every planar graph can be colored with m or fewer colors in such a way that faces sharing a border have different colors.

By "faces sharing a border" we mean faces having one or more edges in common. If two faces have only a vertex or vertices in common we will allow them to have the same color. This is analogous, for example, to maps of the United States in which Utah and New Mexico have the same color.

Notice that this problem is different from the one we have considered previously in this chapter, for now we are coloring the faces, not the vertices, of planar graphs.

The Map Coloring Problem can almost be solved, and the "almost-solution" is that either $m = 4$ or $m = 5$. We shall now derive this.

Let G be a planar graph. Select a point within each face of G and let two such points be connected by a line whenever the corresponding faces of G share a border. As we have seen in Exercise 8 of Chapter 5, this system of points and lines constitutes a graph, called a *dual graph* of G, which is always planar and connected. Figure 124a shows a planar graph G with a dual drawn within it; the dual has been redrawn in Figure 124b.

Suppose that the faces of G are colored in such a way that faces sharing a border have different colors. If "D" denotes a graph dual to G and each vertex of D is given the color of the corresponding face of G, the result is a coloring of D in the sense of Definition

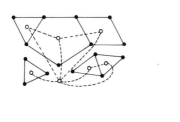

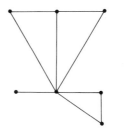

a)

b)

Figure 124

27. Conversely a coloring of the vertices of D in the sense of Definition 27 induces a face-coloring of G in the sense that faces sharing a border have different colors.

Thus the problem of coloring the faces of planar graphs G reduces to the problem of coloring the vertices of the corresponding planar graphs D. Since the Five Color Theorem guarantees that five colors are enough to color the vertices of every planar graph, we know that the number m exists and in fact $m \le 5$. Also, $m \ge 4$ because there are planar graphs, for instance K_4, that require four colors in order that faces sharing a border have different colors. Thus we have a partial solution to the Map Coloring Problem:

If the Four Color Conjecture is true, $m = 4$; if not, $m = 5$.

Hence a mapmaker's paint box need contain only five colors. The mapmaker can rest assured that he or she will always be able to cope, no matter how crazily political boundaries are redrawn.

Opposed to our somewhat idealized mapmakers there are real mapmakers, who work under at least two added restrictions. First, when a country is in two or more pieces like the United States the same color is used for each piece. Second, only oceans and lakes are colored blue. In our statement of the Map Coloring Problem we omitted both restrictions for the sake of simplicity. Including them makes the problem considerably more complex. For example, Figure 125a depicts an island on which there are five countries, one of them in two pieces. By the first restriction, both pieces of Ropia must have the same color, so the island requires five colors in all, and because of the second restriction a sixth must be used for the ocean (see Figure 125b). So our conclusion that a mapmaker's paint box needs only five colors is correct only for the somewhat unorthodox mapmakers who abide by neither restriction.

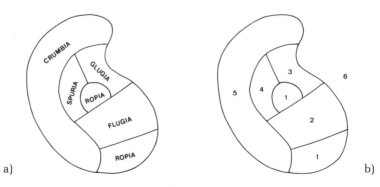

Figure 125

Incidentally, when the Four Color Conjecture was first stated in 1852 it was stated in terms of coloring the faces of planar, in fact polygonal, graphs. See the introduction to Ore's *The Four-Color Problem*.

Exercises

1. Find X for each of the following graphs: C_v, W_v (see Chapter 2, Exercise 6), *UG*, octahedron, dodecahedron, icosahedron. As I remarked on page 136, the icosahedron is a "test case" for the Four Color Conjecture.
2. Find X for each of the graphs in Figure 126.

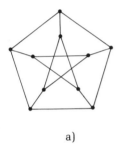

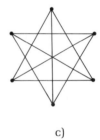

a)   b)   c)

Figure 126

3. Obviously the set of all graphs with $X = 1$ is equal to the set of all null graphs N_v. We might call this the "One Color Theorem". Prove the Two Color Theorem: the set of all graphs with $X = 2$ is equal to the set of all graphs having at least one edge and no odd cyclic subgraphs. (Note: a cyclic graph C_v is called "odd" if v is an odd number and "even" if v is an even number.) If only there were a "Three Color Theorem" the Four Color Conjecture would be practically resolved. For this reason much of today's four-color research is directed toward finding a characterization of graphs with $X = 3$ analogous to the Two Color Theorem.
4. K_3 has $X = 3$, so every supergraph of K_3 has $X \geq 3$. On the other hand C_5 is a graph with $X = 3$ that is not a supergraph of K_3. Find a graph with $X = 4$ that is not a supergraph of K_3.
5. If a graph G has chromatic number X and its complement $\bar{G}$ has chromatic number $\bar{X}$, prove that $X\bar{X} \geq v$. Then use the fact that

$(1/2)(m + n) \geq \sqrt{mn}$ whenever m and n are positive integers to prove that $X + \bar{X} \geq 2\sqrt{v}$.

6. Find a graph for which $X\bar{X} = v$, and a graph for which $X + \bar{X} = 2\sqrt{v}$. Could a single graph satisfy both equations? (If a graph has chromatic number X then "$\bar{X}$" denotes the chromatic number of its complement.)

7. Obviously the Four Color Conjecture is true for planar graphs with four vertices or less. Of the graphs with five vertices only the complete graph K_5 has $X = 5$; K_5 is nonplanar so the Four Color Conjecture is true for planar graphs with five vertices. To prove that the Four Color Conjecture is true for planar graphs with six vertices we argue as follows: K_6 is regular of degree 5, but is nonplanar; every other graph with six vertices has at least one vertex of degree ≤ 4; in particular, therefore, every planar graph with six vertices has at least one vertex of degree ≤ 4, and so has $X \leq 4$ by Theorem 17. Prove that the Four Color Conjecture is true for planar graphs with seven vertices.

8. Color the faces of Figure 123b with four colors. Don't forget the infinite face.

Suggested reading

On the Four Color Conjecture

All books listed at the end of chapter 4 under the heading "On Topology". Also:

The Enjoyment of Mathematics: Selections from Mathematics for the Amateur by Hans Rademacher and Otto Toeplitz (Princeton University Press, 1957), chapter 12.

*"The Island of Five Colors" by Martin Gardner, reprinted in *Fantasia Mathematica* edited by Clifton Fadiman (Simon and Schuster, 1958). Science fiction.

The Four-Color Problem by Oystein Ore (Academic Press, 1967), the Introduction.

"Thirteen Colorful Variations on Guthrie's Four-Color Conjecture" by Thomas L. Saaty in the January, 1972 issue of *The American Mathematical Monthly*.

7. THE GENUS OF A GRAPH

Introduction

This chapter generalizes Chapters 3-6. It is more conceptually difficult and concisely written than the other chapters in this book, so you may find it heavy going, especially if you read it without outside help. Naturally I hope you stick with it, because I think the material is fascinating and so well worth the effort; but as nothing in Chapter 7 will be needed in Chapter 8, you may want to skip Chapter 7 and go on to Chapter 8. Should you decide to do that, however, in order that you will have at least some idea of what Chapter 7 is about, I recommend that you read the next section carefully and at least skim the remainder of the chapter.

The genus of a graph

The term "planar" comprises two conditions. A graph is planar if 1) it is drawn without edge-crossings and 2) it is drawn in a plane. The concept "genus" includes the first condition but generalizes the second by considering graphs drawn on other surfaces.

Figure 127 depicts the first four members of an infinite family of surfaces. These surfaces are taken to be hollow and of negligible thickness. That is, you should think of S_0 as a beachball rather than a baseball, of S_1 as a valveless inner-tube rather than a doughnut, etc. S_0 is called a *sphere*, S_1 is called a *one-hole torus*, S_2 a *two-hole torus*, S_3 a *three-hole torus*, etc. The names of the surfaces are easy to remember if you think of the subscript as referring to the number of holes. Thus S_{79}, say, is a seventy-nine-hole torus and S_0 can be considered a "torus" with no holes.

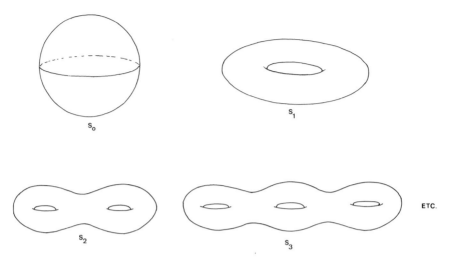

Figure 127

Definition 29. The *genus* of a graph, denoted "*g*", is the subscript of the first surface among the family S_0, S_1, S_2, ..., on which the graph can be drawn without edge-crossings.

Implicit in the definition is the assumption that every graph *has* a genus; i.e. given an immensely complicated and wildly nonplanar graph G, that a systematic search of the sequence S_0, S_1, S_2, ... will eventually reveal *some* surfaces on which G can be drawn without edge-crossings (then g is just the subscript of the first such surface). Soon we will prove that indeed every graph has a genus; but first a theorem and some examples to elucidate this new concept.

Theorem 18. *The set of all planar graphs is equal to the set of all graphs with $g = 0$.*

Proof. To prove that two sets are equal we have to show that each is a subset of the other. Accordingly we have two statements to establish, viz., "Every planar graph has $g = 0$" and "Every graph with $g = 0$ is planar." I'd like to start by proving the latter.

(1)   Every graph with $g = 0$ is planar.

That is, if a graph can be drawn on S_0 without edge-crossings, we have to show that it can also be drawn in a plane without edge-crossings. We do this as follows.

Let G be a graph drawn on a sphere without edge-crossings. Select a point of the sphere which is not a vertex and through which no edges pass, and puncture the sphere at that point. Stretch the hole and gradually flatten the sphere out onto a plane. The result will be G on the plane, still without edge-crossings, and surrounded by a circle (the boundary of the hole). Erase the circle and we have G drawn in a plane without edge-crossings. This process has been illustrated in Figure 128.

The other half of the theorem is

(2)   Every planar graph has $g = 0$.

That is, if a graph can be drawn in a plane without edge-crossings, we have to show that it can also be drawn on S_0 without edge-crossings. We do this by reversing the above procedure.

Let G be a graph drawn in a plane without edge-crossings. Cut out of the plane a circular region containing G and bend this circular region first into a hemisphere and finally into a sphere that is missing one point. Supply the point and the result is a drawing of G on S_0 without edge-crossings. To visualize this you should examine the drawings of Figure 128 in reverse order.

This theorem shows that the concept "planar" is merely a special case of the more general concept "genus". Planar graphs are the graphs of genus 0.

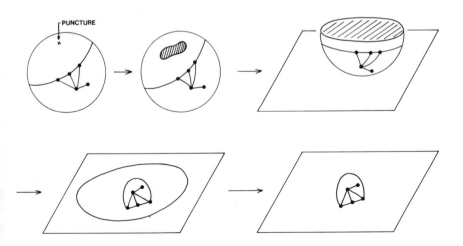

Figure 128

The fact that edges can go "around the back" of a sphere may suggest that edge-crossings can be avoided on spheres when they cannot in a plane, but the theorem demonstrates that this notion is false (see Figure 128). There is, however, a difference between a crossing-free drawing of a planar graph on S_0 and a crossing-free drawing of the same graph in a plane, and that is that the "infinite face" of the plane drawing loses all individuality in the drawing on S_0. There it becomes as finite in extent as all the other faces. Conversely, whichever face of an S_0-drawing of a planar graph is punctured to produce a plane drawing becomes the "infinite face" of the plane drawing. Of course this alteration in no way affects edge-crossings, and it remains true that planar graphs and graphs of genus 0 are precisely the same things.

Examples. 1) K_4 has $g = 0$ because it is planar.

2) Each C_v has $g = 0$ because it is planar.

3) UG has $g = 1$ because it is nonplanar and so cannot, by Theorem 18, be drawn without edge-crossings on S_0; but it can be so drawn on S_1 as Figure 129a shows. Thus 1 is the subscript of the first surface in the family $S_0, S_1, S_2, \ldots$ on which UG can be drawn without edge-crossings.

 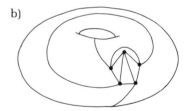

a)   b)

Figure 129

4) K_5 has $g = 1$ for the same reasons as UG. See Figure 129b.

Theorem 19. *Every graph has a genus.*

Proof. Let G be any graph. If G is planar it has $g = 0$ by Theorem 18, so let us assume that G is nonplanar. Take a drawing of G in a plane and transfer the drawing to the surface of S_0 by the procedure used in part (2) of the proof of Theorem 18. Add to S_0 enough tubular "handles" to serve as "overpasses", thereby eliminating the edge-crossings. This has been done in Figure 130 for the case that G is

a)

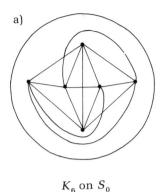

b)

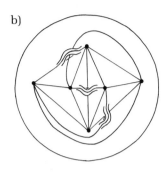

K_6 on S_0 K_6 on S_0-with-three-handles

Figure 130

K_6. Let the number of handles be n. The number n might be quite large but it is finite, as a graph can have only a finite number of edges and hence only a finite number of edge-crossings.

Think of the surface consisting of S_0 with n handles as made of extremely pliable rubber which we are free to stretch, shrink, or otherwise distort as much as we want, provided we don't tear it or fold it back onto itself. With this understanding we see that the surface consisting of S_0 with n handles can be "continuously deformed" (in the sense of pp. 108–109) into S_n, and the edges and vertices of G can be carried along with this continuous deformation. Figure 131 shows this for S_0 with three handles; for the sake of clarity the edges and vertices of K_6 have been omitted.

Thus G can be drawn on S_n without edge-crossings. Since there is at least one member, S_n, of the sequence of surfaces $S_0, S_1, S_2, \ldots$ on which G can be drawn without edge-crossings, there must be a first such member S_g. Then by definition g is the genus of G and we have the theorem.

In Figure 130 three handles are added so $n = 3$ and consequently the genus of K_6 is 1, 2, or 3 (it is not 0 as K_6 is nonplanar). In fact K_6 has $g = 1$ (see Exercise 1).

Theorem 20. *If a graph G has genus g then G can be drawn without edge-crossings on every surface S_n for which $g \leq n$.*

Proof. Take a crossing-free drawing of G and S_g. Find a region on the surface of S_g which contains no edges or vertices of G. The region can be small, no matter. Within this region append to the surface

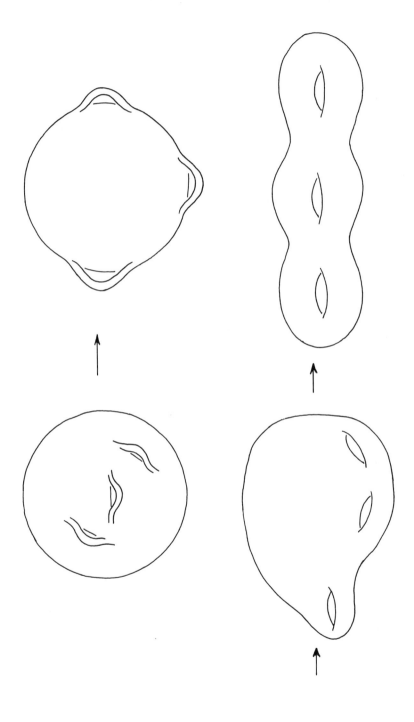

Figure 131

of S_g, $n - g$ tubular handles, where n is any number greater than or equal to g. The resulting surface can be continuously deformed into S_n, with the graph G carried along. The result is a drawing of G on S_n without edge-crossings.

Thus we see that for a specific graph G, its genus g cuts the sequence $S_0, S_1, S_2, \ldots$ into two pieces. The first part $S_0, S_1, \ldots, S_{g-1}$ is finite and consists of all surfaces in the family on which G cannot be drawn without edge-crossings. The second part $S_g, S_{g+1}, \ldots$ is infinite and consists of all surfaces on which G can be drawn without edge-crossings.

Before we defined genus, a graph was simply planar or nonplanar. With the notion of genus we can say much more. If a graph is nonplanar we can find its genus and thereby specify the *extent* to which it is nonplanar. For example a graph of genus 125 is much farther from planarity than a graph of genus 4.

Since planar graphs are precisely the graphs with $g = 0$, the second corollary to Kuratowski's Theorem (Corollary 7a) can be restated: The set of all graphs with $g = 0$ is equal to the set of all graphs that are not supergraphs of expansions of UG or K_5. This suggests that there might be similar theorems to be discovered for graphs of higher genus, i.e. theorems of the form "The set of all graphs with $g = —$ is equal to the set of all graphs that are not supergraphs of expansions of —". So far no such theorems have been discovered. It is known, however, that they exist. Harary reports (*Graph Theory*, p. 117) that in 1968 a mathematician named Vollmerhaus proved that for each genus there are finitely many "forbidden" subgraphs. More precisely Vollmerhaus's result is that for each integer $g \geq 0$ there exist a finite number of graphs $H_1, H_2, \ldots, H_r$ such that the following statement is true: The set of all graphs with genus g is equal to the set of all graphs that are not supergraphs of expansions of H_1 or H_2 or ... or H_r. In the case that $g = 0$, $r = 2$ and the forbidden subgraphs are $H_1 \cong UG$ and $H_2 \cong K_5$. In the other cases $(g > 0)$ it is only known that the graphs H *exist,* that they are *finite* in number, and that their number and structure depend on g—i.e., in general different g's have different r's and different H's. So we know that characterizations of graphs of higher genus, analogous to Kuratowski's for genus 0, exist. But so far none have been stated explicitly and none will be until someone cracks the structure of the H's.

Up to now the term "face" has been defined only for planar graphs. To speak of the "faces" of say, UG, would have been to speak nonsense. But with the new family of surfaces it is now possible to define the term "face" for any graph.

Definition 30. If a graph G of genus g has been drawn on the surface of S_g without edge-crossings, then the edges and vertices of G divide the surface of S_g into regions called the *faces* of G. The number of faces of a graph is denoted "f".

If $g = 0$ this definition reduces to the earlier Definition 23. Note that as before this definition assumes two things: that the edges and vertices of G do in fact cut the surface of S_g into well-defined regions, and that the number of regions is independent of the specific crossing-free drawing of G that is used. Both are true and provable, but we will not demonstrate them.

Examples. 1) UG has $f = 3$. The faces have been numbered in Figure 132a.
     2) K_5 has $f = 5$. See Figure 132b.

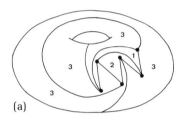
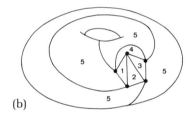

(a)   (b)

Figure 132

A note on counting the faces of nonplanar graphs. In Figure 132a, for example, it seems at first that UG has more than three faces. The confusion is partly due to the drawing, which is a two-dimensional representation of a three-dimensional object. We have to visualize the hidden parts of the torus, which can be difficult. A good rule of thumb is: if you can get from one place to another without leaving the surface, crossing an edge, or passing through a vertex, the two places are in the same face. In Figure 132a it is possible to travel on the surface from any "3" to any other "3" without crossing an edge or passing through a vertex, thus the four "3"'s must lie in the same face. Similar remarks hold for Figure 132b. The unconvinced reader should acquire an inner tube and some chalk.

Euler's Formula says that for planar connected graphs the expression $v + f - e$ always has the value 2. Now that "face" makes sense for nonplanar graphs as well, we might make the hasty conjecture that the same is true for connected graphs of higher genus. That this is not so can be seen by computing $v + f - e$ for the two connected graphs UG and K_5; in each case the value is not 2, but 0. Based on this experiment we might hazard a more sophisticated, though still somewhat daring, conjecture that $v + f - e$ has the same value for connected graphs of the same genus. Incredibly, this is true, and the constant value is related to the genus in a strikingly simple way. The theorem is called Euler's Second Formula and we shall indicate its proof in the next section.

Eulers Second Formula

Euler's Second Formula says that for every connected graph of genus g, $v + f - e = 2 - 2g$. For planar graphs this reduces to the earlier result $v + f - e = 2$, which shall henceforth be known as Euler's First Formula.

Our proof of Euler's Second Formula will not be complete, as it will be based on an assumption that we shall not prove. What follows are some examples to make you feel that the assumption is at least reasonable.

Figure 133a is a drawing of K_5 on S_1. Notice that the walk BCDB goes around through the hole and back again. In Figure 133b the vertices and edges have been rearranged a bit to make a perfect ring out of the walk BCDB.

a) 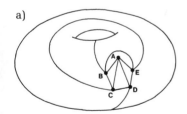   b)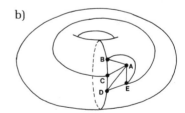

Figure 133

Figure 134a is a drawing of UG on S_1. Again there is a walk— $XAYBX$—that goes around through the hole and can be deformed into a perfect ring (Figure 134b).

The graph of Figure 135 is a less trivial example, having genus 2—see Exercise 6. In Figure 136 the graph has been drawn without edge-crossings on S_2. Notice that there is a walk 2372 going around

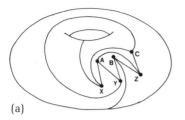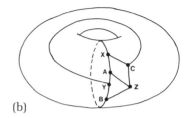

(a)

(b)

Figure 134

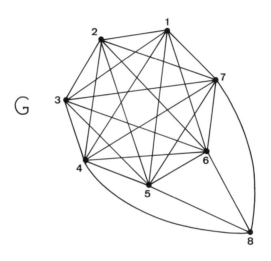

G

Figure 135. A graph of genus 2

through the first hole and another walk 4584 going around through the second hole. In Figure 137 the edges and vertices have been rearranged to make perfect rings out of these walks.

Our assumption is that a similar phenomenon occurs for any connected graph.

Assumption. If G is a connected graph of genus g then there exists a crossing-free drawing of G on S_g such that through each of the g holes of S_g there is a ring composed of vertices and edges of G.

This assumption is reasonable. If G has genus g then G cannot be drawn without edge-crossings on any of the surfaces $S_0, S_1, ...,$ S_{g-1} having fewer than g holes, so each of the g holes is somehow crucial to making a crossing-free drawing of G. Thus at least one

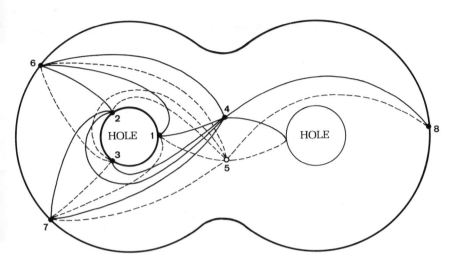

Figure 136. Crossing-free drawing of G on S_2 (view is from the top; vertex 5 and dotted edges are underneath; edges 12, 13, 23, 67, 68, and 78 drawn along rim)

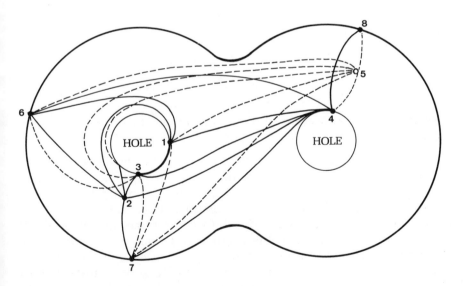

Figure 137. Last drawing rearranged to show rings 2372 and 4584 (vertex 5 and dotted edges are underneath; edges 13, 67, 68, and 78 drawn along rim)

edge of G must pass through each hole, which edge can be joined
with others to make a ring through that hole.

We are now ready to prove Euler's Second Formula. The proof
has been illustrated in Figure 138 for a graph of genus 4. Only the
rings have been drawn for the sake of simplicity. You will have to
imagine the rings as being but a portion of a much larger network
of vertices and edges on the surface S_4.

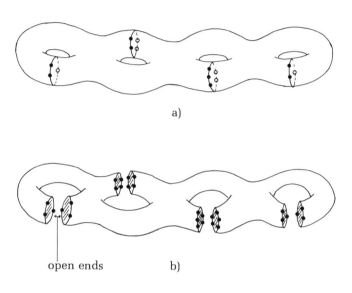

Figure 138

Euler's Second Formula. *If* G *is connected then* $v + f - e = 2 -$
$2g$.

Proof. Let G be a connected graph of genus g. Then by our assumption
there exists a crossing-free drawing of G on S_g such that through
each hole there is a ring composed of vertices and edges of G. Take
a pair of imaginary scissors and cut carefully along each ring, splitting
each vertex and edge of the ring into two vertices and two edges.
The result is that each of the original rings has been replaced by
two new rings, each of which forms the rim of an open-ended tube.
See Figure 138b. Since there were originally g rings there are $2g$
of these rims. Cover these rims with disc-shaped surfaces. The resulting
surface, together with its network of vertices and edges, can be
inflated and then continuously deformed into S_0. See Figure 139.
The final result is a new graph H drawn on S_0. It is clear from the

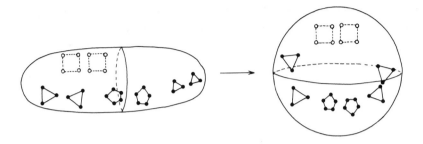

Figure 139

foregoing that the drawing of H on S_0 is crossing-free; it is also true that H is connected—see Exercise 4. Thus H is both planar and connected, Euler's First Formula applies, and we have $v_H + f_H - e_H = 2$. All that remains is to relate the numbers of vertices, faces, and edges of H on S_0 to those of G on S_g.

Let $x = v_H - v_G$. All new vertices were created when we cut the original rings in two lengthwise. Since rings are cyclic graphs and have as many vertices as edges, it follows that $x = e_H - e_G$. Finally, since the only new faces were created when we covered the $2g$ rims, we have $f_H = f_G + 2g$. Thus

$$
\begin{aligned}
v_G + f_G - e_G &= (v_H - x) + (f_H - 2g) - (e_H - x) \\
&= v_H - x + f_H - 2g - e_H + x \\
&= v_H + f_H - e_H - 2g \\
&= 2 - 2g,
\end{aligned}
$$

and the theorem is proved.

Some consequences

Lemma 21. *If a connected graph G has $v \geq 3$ and genus g then $3f \leq 2e$.*

Proof. Imitate the first part of the proof of Theorem 11. I leave the details to you.

Theorem 21. *If G is connected with $v \geq 3$ and genus g then*

$$
g \geq (1/6)e - (1/2)(v - 2).
$$

Proof. By the lemma we know $3f \le 2e$. Since G is connected Euler's Second Formula applies and we have $v + f - e = 2 - 2g$. This can be rewritten $f = -v + e + 2 - 2g$, which upon multiplication by 3 yields $3f = -3v + 3e + 6 - 6g$. Combining this last with the inequality we get $-3v + 3e + 6 - 6g \le 2e$, which can be rewritten

$$-6g \le -e + 3v - 6.$$

Multiplication by $-1/6$ gives the theorem.

Using this theorem we can find a lower bound for the genus of a connected graph, even if we know nothing more than its number of vertices and number of edges.

Example. Let G be a connected graph with 52 vertices and 201 edges. Then by the theorem $g \ge (1/6)201 - (1/2)50 = 33\frac{1}{2} - 25 = 8\frac{1}{2}$. But g is an integer so we can conclude that the genus of G is at least 9.

The next theorem can be used to find an upper bound for the genus of a graph. The theorem was first stated as a conjecture by P. J. Heawood in 1890, but its proof was not completed until 1968. The proof is in two parts and we will do only the easier part. First, two lemmas and a definition. The easy proofs of the lemmas are left to you.

Lemma 22. If a graph H of genus g_H can be drawn on S_n without edge-crossings, then $g_H \le n$.

Lemma 22a. If H is a supergraph of G, then $g_H \ge g_G$.

Definition 31. If x is a number, then "$\{x\}$" denotes the least integer that is greater than or equal to x.

The braces round up to the next whole number, if necessary. If x is an integer then $\{x\} = x$; otherwise $\{x\}$ is the first integer after x.

Examples. $\{3/16\} = 1$; $\{16/3\} = 6$; $\{-7/3\} = -2$; $\{4\} = 4$.

Though "$\{x\}$" also denotes "the set containing x", context should prevent any confusion.

Theorem 22. If $v \ge 3$ the complete graph K_v has genus

$$g = \left\{ \frac{(v-3)(v-4)}{12} \right\}$$

Proof. The idea is to show that

$$g \geq \{(v-3)(v-4)/12\}$$

and

$$g \leq \{(v-3)(v-4)/12\},$$

which together imply the theorem. The proof of the first inequality is based on Theorem 21 and is rather short.

Let v be a positive integer greater than or equal to 3. Then K_v is connected with $v \geq 3$, and $e = (1/2)v(v-1)$ by Theorem 2. Theorem 21 applies and we have

$$g \geq (1/6)e - (1/2)(v-2)$$
$$= (1/6)(1/2)v(v-1) - (1/2)(v-2)$$
$$= (1/12)v(v-1) - (6/12)(v-2)$$
$$= \frac{v(v-1) - 6(v-2)}{12}$$
$$= \frac{v^2 - v - 6v + 12}{12}$$
$$= \frac{v^2 - 7v + 12}{12}$$
$$= \frac{(v-3)(v-4)}{12}.$$

But g is an integer so we can say a bit more:

$$g \geq \{(v-3)(v-4)/12\}.$$

This much Heawood knew. It is the second inequality that is difficult to prove, and whose proof was not completed until 1968. By Lemma 22 the second inequality would be proved if we could draw K_v without edge-crossings on S_n, where

$$n = \{(v-3)(v-4)/12\}.$$

This is exactly what has been done, bit by bit, by various mathematicians since Heawood first conjectured the theorem in 1890. In 1968 the final case was disposed of by G. Ringel and J. W. T. Youngs. Harary recounts the highlights of this 78-year epic in *Graph Theory*, pp. 118–119.

Corollary 22. *If G has $v \geq 3$ and genus g then*

$$g \leq \left\{ \frac{(v - 3)(v - 4)}{12} \right\}.$$

Proof. G is a subgraph of K_v and so by Lemma 22a, g is less than or equal to the genus of K_v.

Corollary 22a. *If G is connected with $v \geq 3$ and genus g then*

$$\{(1/6)e - (1/2)(v - 2)\} \leq g \leq \{(v - 3)(v - 4)/12\}.$$

Proof. Combine Corollary 22, Theorem 21, and the fact that g is an integer.

Estimating the genus of a connected graph

If a connected graph has a relatively large number of edges the last corollary can be used to estimate its genus rather closely. By "relatively large number of edges" we mean that e is close to its maximum value $(1/2)\, v(v - 1)$.

Examples. In the last section we considered a connected graph G with $v = 52$ and $e = 201$. A graph with 52 vertices can have up to $(1/2) \cdot 52 \cdot 51 = 1326$ edges, so G has relatively few edges. The last corollary tells us $\{(1/6) \cdot 201 - (1/2) \cdot 50\} \leq g \leq \{49 \cdot 48/12\}$, or $9 \leq g \leq 196$—not a good estimate.

To see how estimates become better as e becomes larger let us now consider a connected graph H with $v = 52$ and $e = 1200$. 1200 is large relative to the ceiling 1326. This time the lower bound $\{(1/6)e - (1/2)(v - 2)\} = \{(1/6) \cdot 1200 - (1/2) \cdot 50\} = 175$, and so $175 \leq g \leq 196$—a much better approximation.

The best approximation is of course when the upper and lower bounds are the same and g is determined exactly. Let us investigate how close e would have to be to $(1/2)\, v(v - 1)$ for this to happen.

Let "L" denote

$$(1/6)e - (1/2)(v - 2)$$

and "U" denote

$$(v - 3)(v - 4)/12.$$

If $\{L\} = \{U\}$ then either $L = U$ or $0 < U - L < 1$. A graph for which $L = U$ is a complete graph—see Exercise 9—and its genus is known by Theorem 22, so we are uninterested in this case. Suppose then that $0 < U - L < 1$. Unfortunately, it does not follow that $\{L\} = \{U\}$, for no matter how small $U - L$ is, there is always the possibility that there is an integer between L and U, in which case $\{L\}$ is that integer but $\{U\}$ is the next integer. But at least we can say this: if $0 < U - L < 1$ then either $\{L\} = \{U\} - 1$ or $\{L\} = \{U\}$. Let us therefore rearrange the inequality $0 < U - L < 1$ and see what restriction it places on e.

$$0 < U - L < 1$$
$$0 < (v - 3)(v - 4)/12 - (1/6)e + (1/2)(v - 2) < 1$$
$$0 < v^2 - 7v + 12 - 2e + 6v - 12 < 12$$
$$0 < v^2 - v - 2e < 12$$
$$0 > 2e - v^2 + v > - 12$$
$$1/2\, v(v - 1) > e > 1/2\, v(v - 1) - 6$$

We have just proved the following theorem.

Theorem 23. *If G is a connected incomplete graph with $v \geq 3$ and $e \geq (1/2)\, v(v - 1) - 5$, then the upper and lower bounds for g given in the last corollary are either the same integer or consecutive integers.*

Examples. 1) For connected incomplete graphs with $v = 52$ the condition is that $e \geq (1/2) \cdot 52 \cdot 51 - 5 = 1321$, and it so happens that this is enough to determine g in all cases. See the following table.

v	e	L	U	$\{L\}$	$\{U\}$	g
52	1321	195-1/6	196	196	196	196
52	1322	195-1/3	196	196	196	196
52	1323	195-1/2	196	196	196	196
52	1324	195-2/3	196	196	196	196
52	1325	195-5/6	196	196	196	196

2) On the other hand, for connected incomplete graphs with $v = 53$ the condition $e \geq (1/2) \cdot 53 \cdot 52 - 5 = 1373$ is never sufficient to determine g. See the next table.

v	e	L	U	$\{L\}$	$\{U\}$	g
53	1373	203-1/3	204-1/6	204	205	204 or 205
53	1374	203-1/2	204-1/6	204	205	204 or 205
53	1375	203-2/3	204-1/6	204	205	204 or 205
53	1376	204-5/6	204-1/6	204	205	204 or 205
53	1377	204	204-1/6	204	205	204 or 205

g-Platonic graphs

Thus far we have generalized the notion of planarity, introduced in Chapter 3, and have proved a generalization of Euler's First Formula, introduced in Chapter 4. In this and the next section we shall generalize the results of Chapters 5 and 6.

Definition 32. A graph is g-*platonic* if it is a connected, regular graph of genus g such that every edge borders two faces and every face is bounded by the same number of edges.

Specializing the definition to $g = 0$ we have the notion "0-platonic" which coincides with the notion "platonic" defined in Chapter 5.

Notation. In reference to a g-platonic graph, "d" denotes the degree of each vertex, "n" denotes the number of edges bounding each face, and "$P(d, n)$" stands for the quotient

$$\frac{(4g - 4)n}{nd - 2d - 2n}.$$

Lemma 24. If G is g-platonic then $e = dv/2$ and $f = dv/n$.

Proof. Left to the reader.

Lemma 24a. If $g > 0$ and G is g-platonic then $d \geq 3$ and $n \geq 3$.

Proof. Left to the reader.

Lemma 24b. Let $g > 1$. If $(n - 2)(d - 2) > 4$, then $P(d, n) > 0$. Conversely if $P(d, n) > 0$, then $(n - 2)(d - 2) > 4$.

Proof. Left to the reader.

Lemma 24c. *If $g > 1$, $(n - 2)(d - 2) > 4$, $d' \geq d \geq 3$, and $n' \geq n \geq 3$, then $P(d', n') \leq P(d, n)$.*

Proof.

$$(n' - 2)d' \geq (n' - 2)d$$

$$(n' - 2)d' - 2n' \geq (n' - 2)d - 2n'$$

$$n'd' - 2d' - 2n' \geq n'd - 2d - 2n'.$$

$P(d', n')$ and $P(d, n')$ are positive fractions by the previous lemma. They have the same numerator, which is positive since $g > 1$. We have just seen that $P(d', n')$ has the larger denominator, so

(1) $$P(d', n') \leq P(d, n').$$

Now let $x = n'/n$, so $n' = nx$ and we can express $P(d, n')$ as

(2) $$P(d, n') = \frac{(4g - 4)nx}{nxd - 2d - 2nx}.$$

Similarly, since

$$P(d, n) = \frac{(4g - 4)n}{nd - 2d - 2n}$$

$$= \frac{x(4g - 4)n}{x(nd - 2d - 2n)}$$

$$= \frac{(4g - 4)nx}{nxd - 2dx - 2nx},$$

we can express $P(d, n)$ as

(3) $$P(d, n) = \frac{(4g - 4)nx}{nxd - 2dx - 2nx}.$$

So we have only to relate the fractions in equations (2) and (3). They are both positive fractions by Lemma 24b. They have the same numerator, which is positive since $g > 1$. That the first has the larger denominator is a simple consequence of the fact that $n' \geq n$ and so $x \geq 1$:

$$2dx \geq 2d$$

$$-2dx \leq -2d$$

$$nxd - 2dx - 2nx \leq nxd - 2d - 2nx.$$

Therefore the fraction in equation (2) is less than or equal to the fraction in equation (3) and we have

(4) $$P(d, n') \leq P(d, n).$$

If we combine inequalities (1) and (4) the lemma is proved.

In Chapter 5 we discovered that there are infinitely many 0-platonic graphs, all but five of which are either trivial (K_1) or uninteresting (the cyclic graphs). The next theorem says that if $g > 1$ there are at most a finite number of g-platonic graphs. "At most a finite number" means that there are either none at all or, if some exist, that they are finite in number.

What can be easily proved about 1-platonic graphs is less spectacular and is contained in a subsequent theorem.

Theorem 24. *For each $g > 1$ there are at most a finite number of g-platonic graphs.*

Proof. Pick an integer $g > 1$. If for the g selected there are no g-platonic graphs then the theorem is true, so we may as well assume that g-platonic graphs do exist.

Let G be a g-platonic graph. Then $e = dv/2$, $f = dv/n$—both by Lemma 24—and $v + f - e = 2 - 2g$ by Euler's Second Formula. Thus

$$v + dv/n - dv/2 = 2 - 2g$$

$$2nv + 2dv - ndv = 4n - 4ng$$

$$v(2n + 2d - nd) = (4 - 4g)n$$

$$v(nd - 2d - 2n) = (4g - 4)n$$

$$v = \frac{(4g - 4)n}{nd - 2d - 2n} = P(d, n).$$

v is positive so $P(d, n)$ is positive and $(n - 2)(d - 2) > 4$ by Lemma 24b. We also know that $d \geq 3$ and $n \geq 3$ by Lemma 24a.

Case 1: n = 3.

Then $(3 - 2)(d - 2) > 4$ implies that $d \geq 7$, so by Lemma 24c $v = P(d, n) \leq P(7, 3)$.

Case 2: n = 4.

Then $(4 - 2)(d - 2) > 4$ implies that $d \geq 5$, so by Lemma 24c $v = P(d, n) \leq P(5, 4)$.

Case 3: n = 5.

Then $(5 - 2)(d - 2) > 4$ implies that $d \geq 4$, so by Lemma 24c $v = P(d, n) \leq P(4, 5)$.

Case 4: n = 6.

Then $(6 - 2)(d - 2) > 4$ implies that $d \geq 4$, so by Lemma 24c $v = P(d, n) \leq P(4, 6)$.

Case 5: n ≥ 7.

Then any $d \geq 3$ is admissible and $v = P(d, n) \leq P(3, 7)$ by Lemma 24c.

It so happens that $P(3, 7)$ is the largest number among $P(7, 3)$, $P(5, 4)$, $P(4, 5)$, $P(4, 6)$, and $P(3, 7)$ regardless of what number $g > 1$ was chosen at the beginning of the proof—see Exercise 10. Thus in all cases $v \leq P(3, 7)$.

Our assumption that we were dealing with a g-platonic graph G has led to the conclusion that G has at most $P(3, 7)$ vertices. As for each value of $g > 1$ there are only a finite number of graphs with at most $P(3, 7)$ vertices, it follows that for each value of $g > 1$ there are at most a finite number of g-platonic graphs.

Examples. 1) Let $g = 2$. Then

$$P(3, 7) = \frac{(4 \cdot 2 - 4)7}{7 \cdot 3 - 2 \cdot 3 - 2 \cdot 7} = 28,$$

and all 2-platonic graphs can be found among the graphs with $v \leq 28$.

2) Let $g = 3$. Then

$$P(3, 7) = \frac{(4 \cdot 3 - 4)7}{7 \cdot 3 - 2 \cdot 3 - 2 \cdot 7} = 56,$$

and all 3-platonic graphs can be found among the graphs with $v \leq 56$.

3) Let $g = 100$. Then

$$P(3, 7) = (4 \cdot 100 - 4)7/1 = 2772,$$

and all 100-platonic graphs can be found among the graphs with $v \leq 2772$.

Theorem 25. *All* 1-*platonic graphs have either* $d = 3$ *and* $n = 6$, *or* $d = 4$ *and* $n = 4$, *or* $d = 6$ *and* $n = 3$.

Proof. Let G be a 1-platonic graph. Such things do exist—see the examples below. As above we know that $d \geq 3$, $n \geq 3$, $e = dv/2$, $f = dv/n$, and $v + f - e = 2 - 2g$, so

$$v + dv/n - dv/2 = 2 - 2g = 0$$

$$2nv + 2dv - ndv = 0$$

$$v(2n + 2d - nd) = 0.$$

Being the number of vertices of a graph, $v \geq 1$, hence

$$2n + 2d - nd = 0$$

$$nd - 2d - 2n = 0$$

$$nd - 2d - 2n + 4 = 4$$

$$(n - 2)(d - 2) = 4.$$

There are only three combinations of $d \geq 3$ and $n \geq 3$ that satisfy this last equation, and they are the three mentioned in the theorem.

Examples. There exist 1-platonic graphs having each of the three possible combinations. The graph L of Figure 140 is 1-platonic with $d = 3$ and $n = 6$—see Exercise 16. The complete graph K_7 is 1-platonic with $d = 6$ and $n = 3$—see Exercise 17. And the graph Q of Figure 141a is 1-platonic with $d = n = 4$, which we shall now verify.

It is apparent from the drawing that Q is connected and regular of degree 4. We shall show that Q is 1-platonic by showing that it satisfies the other conditions as well.

First, observe that Q is nonplanar because it is a supergraph of an expansion of UG—see Figure 141b and c). Therefore Q has $g \geq 1$. But Q has been drawn on S_1 without edge-crossings in Figure 142, so the genus of Q is exactly 1.

All that remains is to verify that each edge of Q borders two faces and that every face of Q is bounded by 4 edges. This can be done by studying Figure 142. Thus Q is 1-platonic with $d = n = 4$.

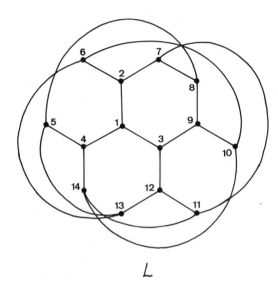

Figure 140. A 1-platonic graph with $d = 3$ and $n = 6$

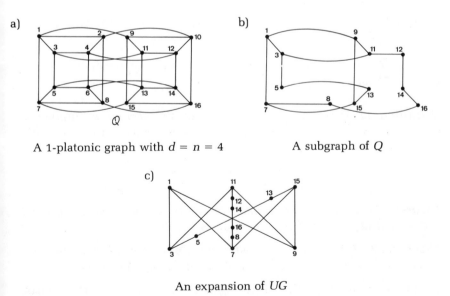

a)

A 1-platonic graph with $d = n = 4$

b)

A subgraph of Q

c)

An expansion of UG

Figure 141.

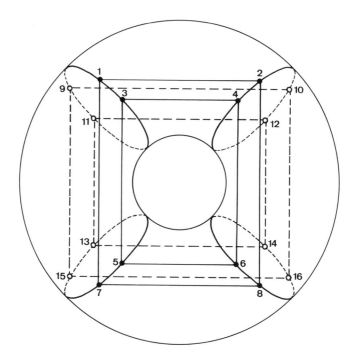

Figure 142. Q on S_1 without edge-crossings (top view; vertices 9–16 and dotted edges are underneath).

The Heawood Coloring Theorem

Definition 33. If $n \geq 0$ is an integer then the *chromatic number of the surface* S_n, *denoted "$X(S_n)$"*, is the largest among the chromatic numbers of graphs having $g \leq n$.

At first this definition can be difficult to grasp. To insure understanding the reader should prove the following two theorems.

Theorem 26. $X(S_n)$ *is the number of colors that is always sufficient and sometimes required to color graphs with $g \leq n$.*

Theorem 27. $X(S_n)$ *is the smallest number of colors sufficient to color every graph that can be drawn on S_n without edge-crossings.*

Implicit in the definition of $X(S_n)$ is the assumption that it exists; i.e. that for every $n \geq 0$ there *is* a largest X for graphs with $g \leq n$. That this is true is a consequence of the Heawood Coloring Theorem, which we shall get to shortly. In the meantime it is easy to prove that at least one surface does indeed have a chromatic number.

Theorem 28. $X(S_0)$ *is either 4 or 5.*

Proof. The Five Color Theorem asserts the existence of a number of colors—namely 5—which is sufficient to color every graph with $g = 0$. The existence of such a number implies the existence of a smallest number with the same property, so $X(S_0)$ exists and in fact $X(S_0) \leq 5$.

On the other hand K_4 has $g = 0$ and $X = 4$ so $X(S_0) \geq 4$.

We can use the notion of chromatic number of a surface to restate the Four Color Conjecture very concisely.

The Four Color Conjecture. $X(S_0) = 4$.

Of course, this has never been proved, so the best we can say is that $X(S_0)$ is 4 or 5. As S_0 is in some sense the "simplest" of the surfaces S_n, and not even *its* chromatic number is known exactly, one would expect no more, and probably less, to be known about $X(S_n)$ for $n > 0$. Incredibly, for every $n > 0$, the exact value of $X(S_n)$ is known! The theorem that gives these values is called the Heawood Coloring Theorem, but before we can state it there are some necessary preliminaries.

Definition 34. If x is a number then "$[x]$" denotes the greatest integer that is less than or equal to x.

Examples. $[3/16] = 0$; $[16/3] = 5$; $[-7/3] = -3$; $[4] = 4$.

Do not confuse $[x]$ with $\{x\}$. They have the same value if x is an integer, but otherwise $[x]$ is the first integer "before" x whereas $\{x\}$ is the first integer "after" x.

Lemma 29. *If x and y are numbers and m is an integer, then*

(1) $$[x] - m = [x - m]$$

and

(2) $$[x][y] \leq [xy].$$

Proof. Let $x = p + i$ and $y = q + j$ where p and q are integers, $0 \leq i < 1$, and $0 \leq j < 1$. The details are left to the reader.

Heawood Coloring Theorem. *If $n > 0$ then*

$$X(S_n) = \left[\frac{7 + \sqrt{1 + 48n}}{2}\right].$$

Proof. Let "t" denote the integer $[(7 + \sqrt{1 + 48n})/2]$. We would have the theorem if we could show that $X(S_n) \le t$ and $X(S_n) \ge t$.

In 1890 Heawood conjectured the theorem and proved the first inequality. One proves the first inequality by showing that t colors are sufficient to color every graph with $g \le n$; it follows that for each $n > 0$, $X(S_n)$ exists and in fact $X(S_n) \le t$. We will not do this, as such a proof would be quite long. The interested reader can find a proof of the first inequality in *Graph Theory* by Harary, pp. 136–137.

The theorem has the hypothesis "$n > 0$" because all known proofs of the first inequality fail for $n = 0$.

The second inequality, which was beyond Heawood's ability, is nowadays quite easy to prove. It follows directly from Theorem 22, proved in 1968.

To prove that $X(S_n) \ge t$ it is sufficient to find a graph with $g \le n$ and $X = t$. It so happens that the complete graph K_t is such a graph. That K_t has $X = t$ is obvious; we will now verify that it has $g \le n$.

For every $n > 0$, $t \ge 3$ so Theorem 22 applies. Thus K_t has genus

$$g = \left\{\frac{(t - 3)(t - 4)}{12}\right\}$$

$$= \left\{\frac{\left(\left[\dfrac{7 + \sqrt{1 + 48n}}{2}\right] - 3\right)\left(\left[\dfrac{7 + \sqrt{1 + 48n}}{2}\right] - 4\right)}{12}\right\}$$

$$= \left\{\frac{\left[\dfrac{7 + \sqrt{1 + 48n}}{2} - 3\right]\left[\dfrac{7 + \sqrt{1 + 48n}}{2} - 4\right]}{12}\right\}$$

(by Lemma 29(1))

$$= \left\{\frac{\left[\dfrac{1 + \sqrt{1 + 48n}}{2}\right]\left[\dfrac{-1 + \sqrt{1 + 48n}}{2}\right]}{12}\right\}$$

$$\leq \left\{ \frac{\left[\dfrac{-1 + \sqrt{1 + 48n}}{4} \right]}{12} \right\} \quad \text{(by Lemma 29 (2))}$$

$$= \left\{ \frac{[12n]}{12} \right\}$$

$$= \left\{ \frac{12n}{12} \right\}$$

$$= \{n\}$$

$$= n.$$

The Heawood Coloring Theorem is an amazing theorem. It tells us that our intuition is wrong in considering S_0 the "simplest" of the surfaces S_0, S_1, S_2, ..., for it is the *only* surface in the family whose chromatic number is not known exactly. Thus from the perspective of coloring at least, S_0 is the most "complicated" surface.

We have mentioned that the proof of the theorem is valid only for $n > 0$, so we can't substitute $n = 0$ into the formula

$$X(S_n) = \left[\frac{7 + \sqrt{1 + 48n}}{2} \right].$$

If we do it anyway—just to see what happens—the result is

$$X(S_0) = \left[\frac{7 + \sqrt{1 + 48 \cdot 0}}{2} \right] = [8/2] = 4,$$

precisely the value claimed by the Four Color Conjecture. Some believers in the Four Color Conjecture derive tremendous psychological support from this. They reason that a "neat" formula like

$$X(S_n) = \left[\frac{7 + \sqrt{1 + 48n}}{2} \right],$$

proved for all positive integers n without exception, must certainly be true for $n = 0$ as well, and that someday the proof of the theorem will be extended to cover this case. Other, more cynical, mathematicians point to Theorem 22 as an example of a "neat" formula that cannot be extended to other cases—see Exercise 5—and suggest that the

Heawood Coloring Theorem might be similar and the Four Color Conjecture false. Perhaps time will tell.* In the meantime the Heawood Coloring Theorem continues to tantalize, insuring a future supply of Four Color addicts.

Exercises

1. Make a crossing-free drawing of K_6 on S_1 and count its faces. Use Euler's Second Formula to verify your count.
2. Make a crossing-free drawing of K_7 on S_1 and count its faces. Use Euler's Second Formula to verify your count.
3. Make a crossing-free drawing of K_8 on S_2.
4. Prove that the graph H mentioned in the proof of Euler's Second Formula is connected.
5. Show that Theorem 22 is false for $v = 1$ and $v = 2$.
6. Prove that the graph of Figure 135 has genus 2.
7. Make a table showing the values of v, f, e, and g for K_v, $1 \le v \le 20$. For what values of v does v exceed g?
8. If G is a graph with 994 vertices and 492,753 edges, estimate its genus.
9. Prove that a graph for which $L = U$ is a complete graph. "L" and "U" are defined on pages 156–157.
10. Prove that for any integer $g > 1$, $P(3, 7)$ is larger than $P(7, 3)$, $P(5, 4)$, $P(4, 5)$, and $P(4, 6)$. "$P(d, n)$" is defined on page 158.
11. **The graph of Figure 126a is nonplanar, so its genus is at least 1. Prove that its genus is exactly 1 by drawing it on S_1 without edge-crossings. Then do the same for the graph of Figure 126b.**
12. Prove that the graph of Figure 143 has genus 3.
13. Find f for the first two graphs in Figure 126, and for the graph in Figure 143.
14. If a graph has 18 edges and 7 vertices, what is its genus? If a graph has 52 edges and 11 vertices, what is its genus?
15. Prove: if a graph is connected of genus g and the boundary of every face is K_3 then $e = 3(v - 2 + 2g)$.
16. Prove that the graph L of Figure 140 is 1-platonic with $d = 3$ and $n = 6$.
17. Prove that K_7 is 1-platonic with $d = 6$ and $n = 3$.
18. The 1-platonic graph Q of Figure 141 is the skeleton of a "toroidal polyhedron", i.e. a polyhedron which, if inflated, would look like S_1. The toroidal polyhedron has been drawn in Figure 144; it

*It has. See Afterword, p. 195.

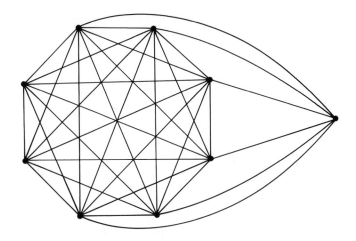

Figure 143

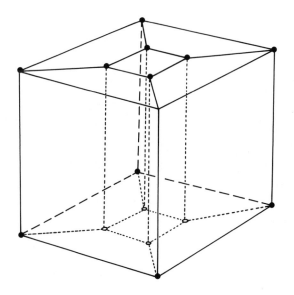

Figure 144

looks like a cube through which a square hole has been bored, causing the top and bottom of the cube to collapse a bit around the hole. K_7 is also the skeleton of a toroidal polyhedron, called the "Császár polyhedron"; read Martin Gardner's article about the

Császár polyhedron in the May, 1975 issue of *Scientific American* and use his pattern to make a cardboard model of it.

19. Is the 1-platonic graph L of Figure 140 also the skeleton of a toroidal polyhedron? If so draw it or make a cardboard model; otherwise explain why not.

20. There are infinitely many 1-platonic graphs. Prove this by describing, and drawing the first few members of, an infinite family of 1-platonic graphs having $d = 4$ and $n = 4$.

21. A graph has $v = 12$ and $X = 10$. Prove that its genus is no less than 4 and no more than 6.

22. **Definition.** The minimum number of edge-crossings with which a graph can be drawn in a plane is called the *crossing number* of the graph, denoted "k".

    Examples. Every planar graph has $k = 0$. K_5 and UG have $k = 1$. K_6 has $k = 3$; it has been drawn with only three edge-crossings in Figure 130a.

    Prove that for every graph $g \le k$.

23. The graph of Figure 104b has $k = 4$. Draw it in a plane with only four edge-crossings. Then prove it has $g = 1$.

24. Find the genus and crossing number of the Petersen graph (Figure 88, page 93).

25. In the course of drawing K_v, where $v \ge 3$, what is the maximum number of edges that can be drawn without an edge-crossing? (For K_5 the answer is 9; for K_6 it is 12.) Though this question can be answered with a simple formula (once you find it, by the way, don't forget to prove that it's correct), no one has yet found a formula for the crossing number k of K_v. This is because there doesn't seem to be any way of predicting how many crossings will be caused when the remaining edges are drawn; for example K_{10} has 45 edges, 24 of which can be drawn without edge-crossings, but the remaining 21 edges cause $k = 60$ crossings.

26. Prove the following analog of Corollary 13 (page 108): if a graph has $g = 1$, then it has at least one vertex with degree ≤ 6.

Suggested reading

On K_7 and the Császár Polyhedron

"Mathematical Games" department by Martin Gardner in the May, 1975 issue of *Scientific American*.

On Crossing Numbers

"Mathematical Games" department by Martin Gardner in the June, 1973 issue of *Scientific American.*

8. EULER WALKS AND HAMILTON WALKS

Introduction

In this chapter we will examine two of the oldest problems in graph theory: "When does a graph have an Euler walk?" and "When does a graph have a Hamilton walk?" Apart from the age of the problems discussed, there is another respect in which this chapter is different from previous chapters. In Chapters 3–7 we were concerned almost exclusively with graphical properties that are defined in terms of the surfaces on which the graphs can be drawn. In this chapter, on the other hand, we will examine graphical properties that are unrelated to the surfaces on which the graphs can be drawn.

Euler walks

Definition 35. A walk $A_1 A_2 \ldots A_{n-1} A_n$ in a graph is *closed* if A_1 and A_n are the same vertex; otherwise the walk is *open*.

Examples. In Figure 145a $DCABCD$, CBC, and $BACBCDCB$ are closed walks; $ACDCACD$, BCA, and $CABCD$ are open walks.

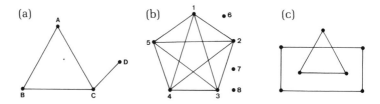

Figure 145

172

Definition 36. An *euler walk* is a walk that uses every edge in the graph exactly once.

Examples. 1) In Figure 145a *DCABC, CABCD,* and *CBACD* are open euler walks; there are no closed euler walks.
     2) Figure 145b has several closed euler walks, one of which is 12345142531; it has no open euler walks.
     3) Figure 145c has no euler walks, open or closed.
     4) *UG* has no euler walks.

It is clear from the definition that a disconnected graph cannot possibly have an euler walk unless all the edges are in one component (see Chapter 4, Exercise 7) as in Figure 145b. Thus we may as well restrict our attention to connected graphs.

Theorem 30. *If a connected graph has a closed euler walk, then every vertex is even. Conversely, if a graph is connected and every vertex is even, then it has a closed euler walk.*

Proof. A vertex is "even" if its degree is an even number—see Exercise 1. Though not difficult, the proof is rather wordy. The theorem is a compound of two statements which we shall prove in turn. The first is

(1)   If a graph is connected and has a closed euler walk then every vertex is even.

This is almost obvious. Let *G* be connected with a closed euler walk that begins and ends at, say, vertex "*A*". Trace the walk with your finger. We start at vertex *A* and leave it along an edge. Since the walk is closed it cannot end at any vertex other than *A*. Thus we will leave every vertex we come to, other than *A*, as many times as we enter it. And since the walk is an euler walk, we will traverse every edge in the graph; so every vertex other than *A* has even degree. Eventually we will enter vertex *A* without being able to leave it—this will be the end of the walk, when we have traversed every edge. Since the walk started at *A*, the degree of *A* is even also.
     If you have trouble visualizing this process I suggest you draw some graphs having closed euler walks and follow along. The second half of the theorem is

(2)   If a graph is connected and every vertex is even then it has a closed euler walk.

This is only slightly more difficult than the first half. Let G be connected with every vertex even. Select any vertex and call it "A". Select any edge incident to A and traverse it. Leave every subsequent vertex on an edge that has not been used, and continue until forced to stop. The result is a walk beginning at A which we can call "W". W doesn't repeat any edges because of the way it was constructed. Figure 146a depicts a connected graph with every vertex even and vertex A chosen. In Figure 146b a walk W has been constructed according to instructions; notice that W is closed. We shall now show that W is necessarily closed.

Whenever we pass through a vertex we use two edges, one when we enter it and one when we leave. Every vertex has even degree, and an even number diminished by 2 is still even; so every time we pass through a vertex there remain an even number of unused edges incident to it, though of course this even number may be 0. Thus every vertex that we have either (1) not touched at all or (2) passed through one or more times has either (1) no unused edges or (2) a positive even number of unused edges. The only vertex to which this does not apply is vertex A, which we touched but did not pass through when we started tracing W; vertex A has an odd number of unused edges. Since every vertex that can be entered can be left, except for vertex A which we will eventually not be able to leave, it follows that when we are forced to stop it will be at vertex A. Thus W is necessarily closed.

It might be that W does not include every edge of G. The walk W of Figure 146b, for example, does not include every edge of the graph. We shall now show that W can be expanded to include any edges left out.

If there are edges of G not included in W let "B" be the first vertex of W having incident edges not in W. Leave B on any unused edge, and leave each subsequent vertex on an unused edge, continuing as long as possible. The result is a walk—call it "X"—that starts at B and doesn't repeat any edges. By the same argument as before, X must be closed—see Figure 146c. Splice X into W and the result is a new walk "Y". Y is a closed walk beginning and ending at A and repeating no edges. If there are edges of G not included in Y, let "C" be the first vertex of Y having edges not in Y and repeat the process. Since graphs have only a finite number of vertices and edges, this can't go on forever, and at some point we will find ourselves in possession of a closed walk, beginning and ending at A, that repeats no edges and uses every edge of G; that is, an euler walk. For the graph of Figure 146 a second splice was necessary—see Figure 146d.

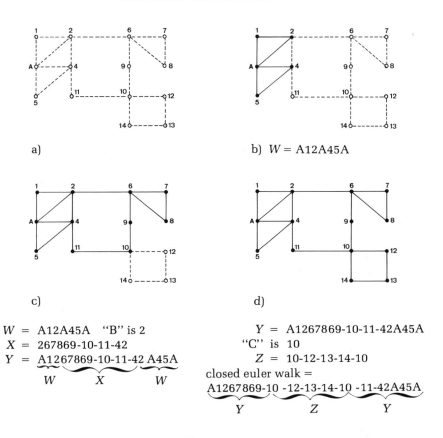

a)

b) $W = A12A45A$

c)

d)

$$W = A12A45A \quad \text{"B" is 2}$$
$$X = 267869\text{-}10\text{-}11\text{-}42$$
$$Y = \underbrace{A12}_{W}\underbrace{67869\text{-}10\text{-}11\text{-}42}_{X}\underbrace{A45A}_{W}$$

$$Y = A1267869\text{-}10\text{-}11\text{-}42A45A$$
$$\text{"C" is } 10$$
$$Z = 10\text{-}12\text{-}13\text{-}14\text{-}10$$
closed euler walk =
$$\underbrace{A1267869\text{-}10}_{Y}\underbrace{\text{-}12\text{-}13\text{-}14\text{-}10}_{Z}\underbrace{\text{-}11\text{-}42A45A}_{Y}$$

Figure 146

This completes the proof of statement (2) and we have the theorem.

Corollary 30. *If a particular vertex is selected from a connected graph having every vertex even, then it is possible to find a closed euler walk beginning and ending at that particular vertex.*

Proof. We proved this while we were proving statement (2) above, because we started with an *arbitrary* vertex A. It could have been any vertex in the graph.

Theorem 31. *If a connected graph has an open euler walk then it has exactly two odd vertices. Conversely, if a connected graph has exactly two odd vertices then it has an open euler walk.*

Proof. A vertex is "odd" if its degree is an odd number (see Exercise 1).

The first half of the theorem is

(1)   If a connected graph has an open euler walk then it has exactly
      two odd vertices.

Let G be connected with an open euler walk that begins at vertex
"A" and ends at vertex "B". Add a new vertex "Z" and two edges
$\{A,Z\}$ and $\{B,Z\}$. The result is a supergraph H of G. Take the open
euler walk A ... B of G, append to it the walk BZA of H, and we
have a closed euler walk A ... BZA in H. By Theorem 30 all vertices
of H are even, so G must have had exactly two odd vertices A and
B.

(2)   If a connected graph has exactly two odd vertices then it has
      an open euler walk.

This is the second half of the theorem. Let G be connected with
exactly two odd vertices "A" and "B". Form a supergraph H of G
by adding a new vertex "Z" and two new edges $\{A,Z\}$ and $\{B,Z\}$.
Then H is a connected graph with all vertices even, so by Corollary
30 H has a closed euler walk beginning and ending at vertex Z. As
Z is adjacent only to A and B, this closed euler walk must look like
this: ZA ... BZ. If we remove the two Z's the result is A ... B, which
must be an open euler walk in G.

This completes the proof of statement (2) and we have the theorem.

Corollary 31. *If a connected graph has an open euler walk, then the
open euler walk must begin at one of the odd vertices and end at
the other.*

Proof. The proof is contained in the proof of part (1) of Theorem
31.

The problem of euler walks is now completely solved. The number
of odd vertices in a graph must be even by Exercise 1. If that even
number is 0 the graph has a closed euler walk; if it is 2 the graph
has an open euler walk; and if it is 4 or more the graph has no
euler walk.

In his interesting book *Mathematics: the Man-Made Universe*, Stein
refers to the problem of euler walks as the "problem of the highway
inspector". A highway system can be viewed as a graph, where
intersections are the vertices. In order to do his job properly, a highway
inspector requires a route taking him over every section of highway
in the system. But in order not to waste time, he would like the

route not to repeat any sections. In short, the highway inspector is looking for an euler walk.

Hamilton walks

Ireland has produced only one famous mathematician, he a bizarre man named William Rowan Hamilton (1805–1865). In 1859, finding himself short of drinking money, he marketed a puzzle called "Around the World". It consisted of a dodecahedron, the vertices of which were labeled with the names of major cities; the task was to discover a route along the edges that would pass through each city exactly once.

Definition 37. An *open hamilton walk* is a walk that uses every vertex in the graph exactly once. A *closed hamilton walk* is a closed walk that uses the initial vertex exactly twice and all the other vertices in the graph exactly once.

Examples. 1) The graph of Figure 147a has several open hamilton walks, two of which are 12435 and 34521. It has no closed hamilton walks.

2) Figure 147b has several closed hamilton walks, two of which are 1234567891 and 7654321987. It has several open hamilton walks,

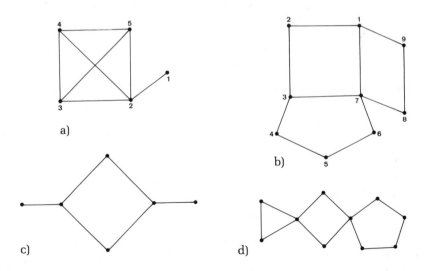

Figure 147

for instance 123456789 and 765432198, which are obtained from the closed hamilton walks by erasing the terminal vertex. Note that every graph with a closed hamilton walk also has an open hamilton walk.

3) The graphs of Figure 147c and d) have no hamilton walks.

Stein calls the problem of hamilton walks the "problem of the traveling salesman". A network of towns and highways can be viewed as a graph; a salesman who has stops to make in each town, but doesn't want to waste time by passing through towns more than once, is naturally interested in finding a hamilton walk through the network.

As defined, euler walks and hamilton walks are very similar things. An euler walk uses each edge exactly once, a hamilton walk uses each vertex exactly once. As the problem of euler walks is completely solved, one might expect the same to be true for hamilton walks. Surprisingly, this is not so. Most of what is known about hamilton walks is contained in two rather uninformative theorems that we shall prove in this section. First a lemma, the proof of which I leave to you.

Lemma 32. *If the sum of the degrees of every pair of vertices of a graph G is at least* $v - 1$, *then*

(1) *every pair of vertices are either adjacent to each other or to a common third vertex, and*

(2) *G is connected.*

Proof outline. Assume that (1) is false, i.e. that there are vertices A and B which are not adjacent to one another or to a common third vertex. Derive from this the conclusion that the degree of A plus the degree of B is less than $v - 1$, contradicting the hypothesis of the theorem. Then (2) follows quickly from (1).

Theorem 32. *If the sum of the degrees of every pair of vertices of a graph G is at least* $v - 1$, *then G has an open hamilton walk.*

Proof. Let w be the largest integer such that the path graph P_w is a subgraph of G. See Chapter 2, Exercise 5 for the definition of "path graph".

Were w to equal v, P_w would be an open hamilton walk in G. We shall therefore essay to show that $w = v$.

Assume for the sake of argument that $w < v$. Our goal is to deduce a contradiction.

Figure 148a illustrates the subgraph P_w of G. Since $w < v$, G has vertices not in the drawing. And since G is connected—which it is by Lemma 32—G also has edges not in the drawing. These you will have to imagine. Let "C" and "D" be the vertices at the ends of

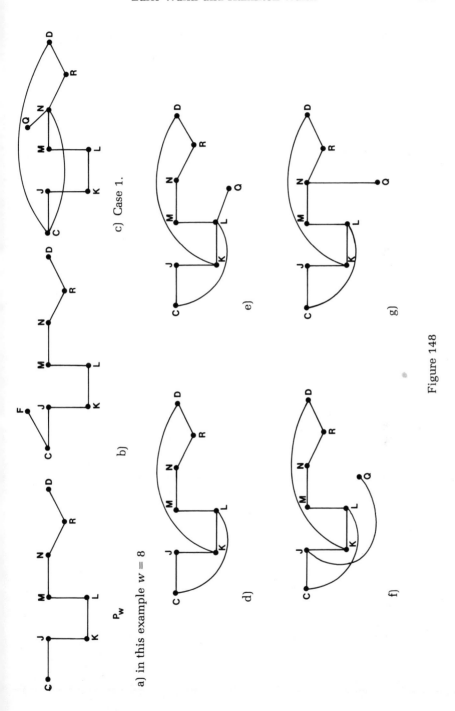

a) in this example $w = 8$

b)

c) Case 1.

d)

e)

f)

g)

Figure 148

P_w. Note that C and D can be adjacent only to other vertices of P_w. For if C, say, were adjacent to some vertex F which is not a vertex of P_w—as in Figure 148b—then $FCJKLMNRD \cong P_{w+1}$ would be a subgraph of G in conflict with the maximality of the integer w.

Case 1. C and D are adjacent.

Let Q be a vertex of G not included in P_w. By the lemma either C and Q are adjacent to each other—which we have just noted is impossible since C is adjacent only to other vertices of P_w—or C and Q are adjacent to a common third vertex. This common third vertex, being adjacent to C, must be in P_w; say it is N, for the sake of example. Then $QNMLKJCDR \cong P_{w+1}$ is a subgraph of G, contradicting the maximality of w. So we have a contradiction for this case.

Case 2. C and D are not adjacent.

For this discussion we shall think of the direction along P_w from C toward D as being "left to right". With this understanding we can say "K is to the left of L" and "M is to the right of L."

Let y be the number of vertices other than R that are adjacent to D, so $y = \deg D - 1$. These y vertices are all vertices of P_w as noted above. We want to show that C is adjacent to one of the y vertices which are immediately to the right of the y vertices adjacent to D. For example if y were 3 and the vertices other than R that are adjacent to D were J, K, and M, we would want to show that C is adjacent to K, L, or N. We do this by contradiction.

Suppose that C is not adjacent to any of the y vertices which are immediately to the right of the y vertices (other than R) adjacent to D. We shall derive an upper bound for $\deg C$. Using first the information that C is adjacent only to other vertices of P_w, we can make the rather coarse statement that $\deg C \leq w - 1$. Using next the information that C and D are not adjacent, we can sharpen this a bit into $\deg C \leq w - 2$. Using finally our supposition that C is not adjacent to any of the y vertices to the right of the y vertices adjacent to D, we can say that in fact $\deg C \leq w - 2 - y$. Now by the definition of y, $\deg D = y + 1$; so $\deg C + \deg D \leq w - 2 - y + y + 1 = w - 1 < v - 1$, which contradicts the hypothesis of the theorem.

So C is adjacent to one of the y vertices immediately to the right of the y vertices—other than R—adjacent to D. This situation has been illustrated in Figure 148d, with D adjacent to K and C adjacent to L.

Let Q be a vertex of G not included in P_w. As before, Q must be adjacent to some vertex of P_w which is in turn adjacent to C. If this vertex is L, as in Figure 148e, then $QLCJKDRNM \cong P_{w+1}$ is a subgraph of G, contradicting the maximality of w. If this vertex is to the left of L, say it is J, then $QJCLKDRNM \cong P_{w+1}$ is a subgraph of G—contradiction. If it is to the right of L, say it is N, then we have the same contradiction with $QNRDKJCLM \cong P_{w+1}$.

So we have contradictions in this case too; thus w is not smaller than v. Hence $w = v$ and the theorem is proved.

Theorem 33. *If the sum of the degrees of every pair of vertices of a graph G is at least v, then G has a closed hamilton walk.*

Proof. This is really a corollary to the previous theorem. Let G be a graph satisfying the hypothesis. Then by the previous theorem G has an open hamilton walk P_v. Figure 149a depicts such a walk. All the vertices of G are in the drawing, since P_v is a hamilton walk; there may be edges left out, which the reader will have to imagine.

Case 1. C and D are adjacent.

Then G has a closed hamilton walk, shown in Figure 149b.

Case 2. C and D are not adjacent.

Then C is adjacent to a vertex immediately to the right of a vertex adjacent to D, for otherwise (as before) $\deg C \leq v - 2 - (\deg D$

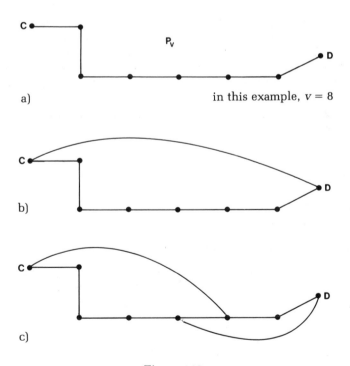

in this example, $v = 8$

Figure 149

$- 1)$ so $\deg C + \deg D \le v - 2 - \deg D + 1 + \deg D = v - 1 < v$—contradiction. Hence the situation is as depicted in Figure 149c and G has a closed hamilton walk.

These two theorems leave much to be desired. First, the hypotheses, though sufficient to guarantee the respective conclusions, are not even remotely necessary. Consider for example the simple graph C_6. Though it obviously has both open and closed hamilton walks, if any two vertices are chosen the sum of their degrees is 4, which is less than $5 = v - 1$ and therefore too small for either theorem to apply. And so though C_6 has both open and closed hamilton walks, we would never know it from the theorems. See Exercise 8 for more examples.

Second, there is no surprise to them. Paraphrased, Theorem 32 says that if a graph has rather a lot of edges, and if these edges are rather evenly distributed around the vertices, then the graph has an open hamilton walk. Theorem 33 says that a graph with even more evenly-distributed edges has a closed hamilton walk. Upon reflection neither statement is startling.

Third, the theorems are hard to apply. To determine if either theorem yields information about a graph with, say, 20 vertices, one would have to compute the degree-sum for each of the 190 possible pairs of vertices. Surely, just by inspecting the graph closely, one could determine if it had hamilton walks in less time than that would take.

There is one respect, however, in which the two theorems are remarkable: they represent practically all that is known about hamilton walks. Though apparently so similar to euler walks, about which everything is known, it seems that hamilton walks are at root much more complicated things.

Multigraphs

Definition 38. A *skein* is an object consisting of two dots connected by two or more lines.

Examples. The first five skeins are shown in Figure 150.

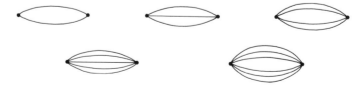

Figure 150

Definition 39. If some of the edges of a graph G, together with their incident vertices, are replaced by skeins, the result is called a *multigraph M(G) generated by G.*

Note that each multigraph has a unique generator, which can be retrieved by erasing all but one of the lines in each skein. On the other hand, a single graph G generates infinitely many multigraphs $M_1(G)$, $M_2(G)$, $M_3(G)$, etc. Note also that every graph is a multigraph since in mathematics the word "some" includes the possibility "none".

Examples. Figure 151 depicts a graph and three multigraphs generated by it.

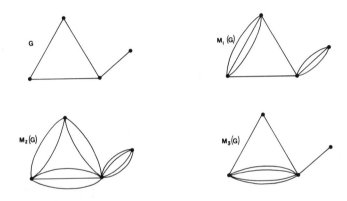

Figure 151

This book could have been about multigraphs. That is, most of the concepts we have discussed—complement, isomorphism, coloring, genus, etc.—could have been defined to include multigraphs and, had this been done, most of the theorems would have remained true. In this section we shall extend the definitions of "walk" and related concepts to include multigraphs, and shall indicate how to re-prove Theorems 30 and 31 for multigraphs. The interested reader might care to do the same for other theorems in this book.

Definition 40. A *walk* in a multigraph is an alternating sequence $A_1 B_1 A_2 B_2 \ldots A_{n-1} B_{n-1} A_n$ of not necessarily distinct vertices and edges, beginning and ending with a vertex, such that each edge connects the vertices flanking it in the sequence. If $A_1 = A_n$ the walk is *closed;* otherwise it is *open.* A multigraph is *connected* if every pair of vertices are joined by a walk; otherwise it is *disconnected.* An *euler walk* is a walk that uses every edge of the multigraph exactly once. In a multigraph the *degree* of a vertex is the number of incident edges. A vertex is *odd* or *even* according to whether its degree is odd or even.

Examples. In Figure 152, multigraph a) has open walks such as A3D7E6D6E and closed walks like A4D8C11B2A; it is connected; it has open euler walks such as B2A5D6E7D4A3D8C9B10C11B1C but no closed euler walks; vertex *D* is even, having degree 6, and vertex *B* is odd, having degree 5. Multigraph b) has open and closed walks too, but being disconnected has no euler walks; it has two odd vertices and four even ones.

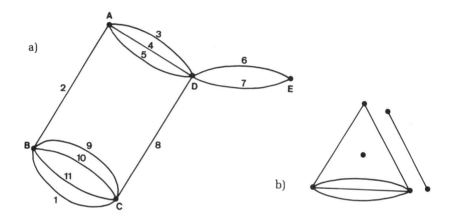

Figure 152

Theorem 34. *If a connected multigraph has a closed euler walk then every vertex is even. Conversely, if a multigraph is connected and every vertex is even, then it has a closed euler walk.*

Proof outline. There are two separate statements to prove. For each, start with a connected multigraph $M(G)$ and splice one vertex of degree 2 into each of its edges. The result is a connected *graph H.* Apply Theorem 30 to H and translate back to $M(G)$. Note the convert-solve-reconvert procedure.

Corollary 34. *If every vertex of a connected multigraph is even and a particular vertex is selected, then it is possible to find a closed euler walk beginning and ending at that particular vertex.*

Theorem 35. *If a connected multigraph has an open euler walk then* *it has exactly two odd vertices. Conversely, if a multigraph is connected and has exactly two odd vertices, then it has an open euler walk.*

Proof outline. We *could* prove this theorem by imitating the proof of Theorem 34, the only difference being that we would apply Theorem 31 instead of Theorem 30. But that would be dull. Here's another approach:

Let $M(G)$ be a connected multigraph with an open euler walk. Add to $M(G)$ a new edge joining the ends of the open euler walk, producing a new multigraph $M(H)$ having a closed euler walk. Apply Theorem 34 to $M(H)$, remove the new edge, and conclude that $M(G)$ has exactly two odd vertices.

To get the other half of the theorem let $M(G)$ be a connected multigraph having exactly two odd vertices. Add a new edge joining the odd vertices, producing a new multigraph $M(H)$ having every vertex even. Apply Corollary 34 to $M(H)$, remove the new edge, and conclude that $M(G)$ has an open euler walk.

Corollary 35. *If a connected multigraph has an open euler walk, the open euler walk begins at one of the two odd vertices and ends at the other.*

The Königsberg Bridge Problem

Recently there was announced a problem that, while it certainly seemed to belong to geometry, was nevertheless so designed that it did not call for the determination of a magnitude, nor could it be solved by quantitative calculation; consequently I did not hesitate to assign it to the geometry of position, especially since the solution required only the consideration of position, calculation being of no use. In this paper I shall give an account of the method that I discovered for solving this type of problem, which may serve as an example of the geometry of position.

Thus runs the birth notice of "the geometry of position", a branch of mathematics now called "topology". The quotation is taken from the first paragraph of Euler's epochal 1736 paper "The Seven Bridges of Königsberg." Euler goes on to describe the problem as follows.

The problem, which I understand is quite well known, is stated as follows: In the town of Königsberg in Prussia there is an island A, called "Kneiphof", with the two branches of the river (Pregel) flowing around it, as shown in figure [153]. There are seven bridges, a, b, c, d, e, f and g, crossing the two branches. The question is whether a person can plan a walk in such a way that he will cross each of these bridges once but not more than once. I was told that while some denied the possibility of doing this and others were in doubt, there were none who maintained that it was actually possible. On the basis of the above I formulated the following very general problem for myself: Given any configuration of the river and the branches into which it may divide, as well as any number of bridges, to determine whether or not it is possible to cross each bridge exactly once.

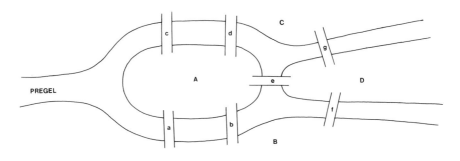

Figure 153

How true to the mathematical spirit is Euler's move to formulate for himself a "very general problem." Solving one specific problem isn't any fun. Given a specific problem, a pure mathematician will set it aside, ignore it, and turn his/her efforts to constructing an enormous machine—made up of definitions, lemmas, theorems, etc.—that will not only solve the original problem, but also a wide range of similar problems. Of course, more time and energy go into the construction of the machine than would have been required to solve merely the original problem. But the pure mathematician's goal is to have a good time, not to be efficient; and machine-building is a lot more fun than problem-solving.

Euler builds his machine by first abstracting from Figure 153 the notion of multigraph. Letting land areas be vertices and bridges be edges, Figure 153 becomes the multigraph of Figure 154, and the original problem "whether a person can plan a walk in such a way that he will cross each of these bridges once but not more than once" becomes "does the multigraph of Figure 154 have an euler walk?"

Euler was not so vain as to christen such a walk an "euler walk". Nor does there appear in his paper the word "multigraph" or any drawing resembling Figure 154. We are attaching modern labels to Euler's thoughts; nonetheless, we are being faithful to them.

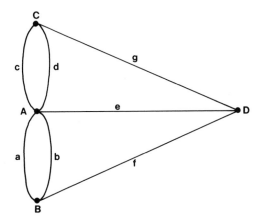

Figure 154

Euler's next step is to prove two theorems.

Theorem 36. *The sum of the degrees of the vertices of a multigraph is 2e.*

Proof. Left to the reader. This theorem is an extension to multigraphs of Chapter 2, Exercise 11.

Theorem 37. *Every multigraph has an even number of odd vertices.*

Proof. Left to the reader. This extends Exercise 1.

Finally he proves Theorems 34 and 35, and their corollaries; and the machine is complete. Here it is in the builder's own words:

> Thus for any configuration that may arise the easiest way of determining whether a single crossing of all the bridges is possible is to apply the following rules:
> If there are more than two regions which are approached by an odd number of bridges, no route satisfying the required conditions can be found.
> If, however, there are only two regions with an odd number of approach bridges the required journey can be completed provided it originates in one of the regions.
> If, finally, there is no region with an odd number of approach bridges, the required journey can be effected, no matter where it begins.
> These rules solve completely the problem initially proposed.

Once the machine is built and stands looming over the puny problem that suggested its construction, the pure mathematician can act out his/her violent fantasies by activating the machine and watching it pulverize the original problem. In our situation, feeding the Königsberg Bridge Problem to Euler's machine produces an immediate solution: No, one cannot take a walk and cross every bridge exactly once because there are four regions touched by an odd number of bridges. Or, in modern terminology, Figure 154 has no euler walks because it has four odd vertices.

Euler's paper is a first-rate example of mathematical thinking. For that reason it is required reading in many college English courses.

In the last section we remarked that this book could have been about multigraphs; one might wonder that it hasn't been, since multigraph theory is apparently older than graph theory. The reason is that the added complexity of multigraphs makes them more difficult to deal with, but there are virtually no interesting multigraph theorems to make the added difficulty worthwhile. A theorem in multigraph theory almost always has an equally interesting counterpart in graph theory.

By the way, Königsberg is not on any map. The town has been renamed "Kaliningrad"; it is in Russia on the Baltic Sea.

Exercises

1. **Definition.** A vertex of a graph is *odd* if its degree is an odd number; otherwise it is *even*.

   Use Exercise 11 of Chapter 2 to prove that every graph has an even number of odd vertices.

2. Find all integers $v \geq 2$ for which
   a) K_v has an open euler walk
   b) K_v has a closed euler walk
   c) K_v has an open hamilton walk
   d) K_v has a closed hamilton walk.

3. Trace each drawing in Figure 155 without lifting your pencil from the paper or going over any lines more than once. a) is a bad puzzle because it can't be done no matter where you start (explain why); b) is a bad puzzle because it *can* be done no matter where you start (explain why); c) is a good puzzle because it can be done, but only if you start at one of two places (explain why).

4. In chess, a "knight's move" consists of two squares either vertically

a)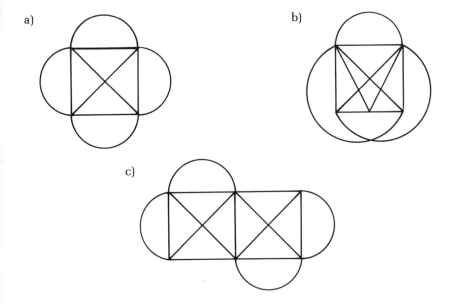

b)

c)

Figure 155

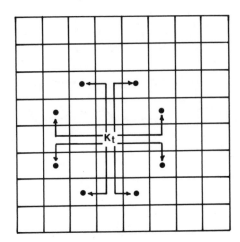

Figure 156

or horizontally and then one square in a perpendicular direction. Depending on where the knight is situated, he has a minimum mobility of two moves—when in a corner—and a maximum mobility of eight moves, as shown in Figure 156. Let C be a graph with $v = 64$, its vertices corresponding to the squares of a chessboard. Let two vertices of C be joined by an edge whenever a knight can go from one of the corresponding squares to the other in one move. Does C have an euler walk? (You don't have to draw C to answer.)

5. Prove that the graph C of the previous exercise has a closed hamilton walk. Such a walk is called a "knight's tour" by puzzle enthusiasts.

6. Figure 157 depicts a system of bridges and land areas. Can you take a walk and cross each bridge exactly once? If so, where do you start and finish? Blow up the bridge from H to I and answer the same two questions.

7. For each integer $v \geq 2$, find a graph with v vertices in which the sum of the degrees of every pair of vertices is at least $v - 1$, but which has no closed hamilton walk. Of course the graph will have an open hamilton walk by Theorem 32.

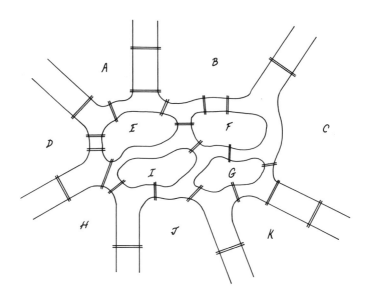

Figure 157

8. Show that every path graph P_v (see Chapter 2, Exercise 5) has an open hamilton walk, though Theorem 32 is applicable only to P_2 and P_3. Then show that every wheel graph W_v (see Chapter 2, Exercise 6) has a closed hamilton walk, though Theorem 33 is applicable only to W_4, W_5, and W_6.

9. Satisfy yourself that every graph with v vertices and a closed hamilton walk is a supergraph of the cyclic graph C_v. Then use this fact to prove that if a graph has a closed hamilton walk then its connectivity c is at least 2 (see Chapter 4, Exercise 11).

10. There are v guests to be seated at a single round table. Each guest is acquainted with at least $\{v/2\}$ of the others. Prove that they can be seated in such a way that each guest is between two acquaintances. (If $v/2$ is not a whole number, the braces round it up to the next whole number.)

11. Prove: if $n \geq 2$ and G is connected with $2n$ odd vertices then G has n open walks which, together, use every edge of G exactly once.

12. Draw four graphs with $v = 8$, the first with an euler walk but no hamilton walk, the second with a hamilton walk but no euler walk, the third with both and the fourth with neither. This indicates that there is no simple relationship between euler walks and hamilton walks. (There is, however, a complicated relationship—see Exercise 18.)

13. **Definition.** If G is a graph with $e \neq 0$ then the *line graph of G*, denoted "$L(G)$", is the graph having one vertex corresponding to each edge of G and such that two vertices of $L(G)$ are joined by an edge whenever the corresponding edges of G share a vertex.

   Examples. 1) Figure 158a is a graph whose line graph has been drawn in Figure 158b. Each vertex of $L(G)$ corresponds to an edge of G. In the figure this correspondence has been made explicit by numbering the edges of G and giving the same numbers to the corresponding vertices of $L(G)$. Two vertices of $L(G)$ are joined whenever the corresponding edges of G share a vertex.

a)

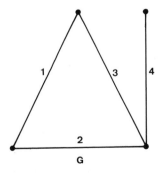

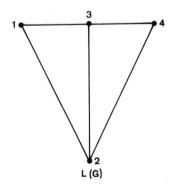

Figure 158

c)

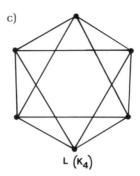

L (K_4)

Figure 158 (continued)

2) Figure 158c is the line graph of K_4. K_4 has six edges, each adjacent to four of the others, thus its line graph has six vertices, each adjacent to four of the others.

Of the eleven graphs with $v = 4$ (drawn in Figure 45), N_4 has no line graph because it has $e = 0$ and we have just drawn the line graph of K_4. Draw the line graphs of the remaining nine.

14. Prove: if G has a closed euler walk then $L(G)$ has both a closed euler walk and a closed hamilton walk.

15. Find a graph G such that $L(G)$ has both a closed euler walk and a closed hamilton walk but G has no closed euler walk. This shows that the converse of Exercise 14 is false.

16. Prove: if G has a closed hamilton walk then $L(G)$ has one too.

17. Find a graph G such that $L(G)$ has a closed hamilton walk but G hasn't. This shows that the converse of Exercise 16 is false.

18. **Definition.** If G is a graph with $e \neq 0$, then the *trisection graph of* G, denoted "$T(G)$", is the expansion of G formed by splicing two vertices of degree 2 into every edge of G.

Example. The graph of Figure 159a is the trisection graph $T(G)$ of the graph G of Figure 158a. Figure 159b shows the line graph $L(T(G))$ of $T(G)$.

Exercises 14 and 16 establish a partial relationship between euler walks and hamilton walks. The following theorem gives a more profound relationship.

Theorem. *A graph G with $e \neq 0$ does or doesn't have a closed euler walk according as $L(T(G))$ does or doesn't have, respectively, a closed hamilton walk.*

a)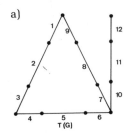

b)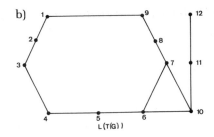

Figure 159

Draw some graphs G with closed euler walks and in each case check that $L(T(G))$ has a closed hamilton walk. Then draw a few graphs G without closed euler walks and in each case check that $L(T(G))$ doesn't have a closed hamilton walk. With these examples under your belt try to prove the theorem.

19. It is usually true that different (i.e., nonisomorphic) connected graphs have different line graphs. In fact there is only one exception: there are two different connected graphs having line graph K_3. Find them.

20. Let G be a graph with v vertices and e edges, and let the degrees of the vertices of G be "d_1", "d_2", ..., and "d_v". Of course $L(G)$ has e vertices. Prove now that the number of edges of $L(G)$ is equal to $(1/2)(d_1^2 + d_2^2 + ... + d_v^2) - e$.

21. Prove: cyclic graphs are isomorphic to their line graphs, and are the only graphs having that property.

Suggested reading

On Euler Walks and Hamilton Walks

*"The Seven Bridges of Königsberg" by Leonhard Euler, reprinted in volume 1 of *The World of Mathematics,* edited by James R. Newman (Simon and Schuster, 1956).

Mathematics: the Man-Made Universe by Sherman K. Stein (W. H. Freeman, 1963), chapter 8.

The Enjoyment of Mathematics: Selections from Mathematics for the Amateur by Hans Rademacher and Otto Toeplitz (Princeton University Press, 1957), chapter 2.

On William Rowan Hamilton

Men of Mathematics by Eric Temple Bell (Simon and Schuster, 1962), chapter 19.

*"William Rowan Hamilton" by Sir Edmund Whittaker in the May, 1954 issue of *Scientific American,* reprinted as chapter 7 of *Mathematics in the Modern World: Readings from Scientific American* with introductions by Morris Kline (W. H. Freeman, 1968).

On the Origins of Graph Theory

Graph Theory by Frank Harary (Addison-Wesley, 1969), chapter 1.

AFTERWORD

As *Dots and Lines* went to press in 1976, yet another "the Four Color Conjecture has been resolved" rumor was circulating. Only this time the rumor was confirmed.

> Toiling in an arcane area that totally baffles most ordinary mortals, mathematicians usually despair of even trying to explain their work to laymen. Yet recently two University of Illinois mathematicians announced a breakthrough of such widespread interest that even the reticent American Mathematical Society issued a rare press release. The news: after more than a century of futile brain racking, one of mathematics' most famous teasers—the so-called four-color conjecture—has finally been proved.
> —"Eureka!", *Time*, Sept. 20, 1976, pp. 87–88

The mathematicians were Kenneth Appel and Wolfgang Haken. The proof is described in two articles in the September, 1977 issue of *Illinois Journal of Mathematics.* Appel and Haken have also written a less technical account for laymen in the October, 1977 issue of *Scientific American.* To make this article more accessible to readers of *Dots and Lines* I would like to recast our proof of the Five Color Theorem (pp. 131–35). Though equivalent logically to the original induction proof, this new proof is rather different psychologically.

The Five Color Theorem. *Every planar graph has X ≤ 5.*

Proof. The proof is indirect. We shall suppose that there are some planar graphs with $X > 5$, and show that this assumption leads to a contradiction.

From among the planar graphs with $X > 5$ that we are supposing to exist, choose one—call it G—having the minimum number of vertices that such graphs ever have. That is, choose a planar graph G so that

(1) G has $X > 5$ and (2) every planar graph with fewer vertices has $X \leq 5$. I want to establish two statements about G, the first of which is

I. None of the graphs in Figure 160 can be a subgraph of G. This is because, if any of the six graphs were a subgraph of G, then $G - A$ (the subgraph of G obtained by erasing vertex A and all edges incident to vertex A) would have $X \leq 5$ (by (2) above) and so G would also have $X \leq 5$ (by Cases 1, 2, and 3 on pp. 132–35). But G has $X > 5$, so none of the graphs in Figure 160 can be a subgraph of G.

Figure 160

To follow the argument of the preceding paragraph you will have to reread, at the point at which they are mentioned, Cases 1, 2, and 3 of the original proof. To avoid confusion do not read any other parts of the original proof. Begin with the words "Case 1" on p. 132 and stop above the figure on p. 135.

The second statement I want to establish about G is

II. At least one of the graphs in Figure 160 must be a subgraph of G.

This is by Corollary 13 (p. 108).

We are in quite a bind. By Statement I *none* of the graphs in Figure 160 can be a subgraph of G, yet by Statement II one of them *must* be. The only way out is to conclude that we were wrong at the outset in supposing that there exist planar graphs with $X > 5$. Thus every planar graph has $X \leq 5$.

In the terminology of Appel and Haken, Statement I says that each of the graphs—they would say "the configurations"—in Figure 160 is "reducible". The word "reducible", which they use for historical reasons only and is not subject to common-sense interpretation, means roughly "cannot occur as a subgraph of the graph under consideration".

Continuing in their terminology, Statement II says that the set of configurations in Figure 160 is "unavoidable", since the group cannot as a whole be avoided by (i.e., at least one must be a subgraph of) the graph under consideration.

Using these terms we can summarize our new proof of the Five Color Theorem as follows: We have shown that Figure 160 is an unavoidable set of reducible configurations for a planar graph G having $X > 5$ and minimal vertices. Finding this set means that no such graph exists, thus there can be no planar graphs with $X > 5$ at all, thus every planar graph has $X \leq 5$.

The Appel-Haken proof of what we should now call the Four Color Theorem has the same organization. To show that every planar graph has $X \leq 4$, they suppose that planar graphs exist with $X = 5$ (i.e. $X > 4$, in view of the Five Color Theorem), and focus their attention on one (a "minimal five-chromatic" planar graph) having the minimum number of vertices that such graphs ever have. They then produce an unavoidable set of 1482 reducible configurations for this hypothetical graph, concluding that no minimal five-chromatic planar graph exists, hence no five-chromatic planar graphs exist and every planar graph has $X \leq 4$.

Appel and Haken begin their *Scientific American* article by considering the Four Color Conjecture in its dual form as a problem about coloring the faces, rather than the vertices, of planar graphs (we did the same in "Coloring Maps", pp. 136–39). But midway through they convert the conjecture into the form in which we first studied it.

In the same article they remark that though this famous conjecture has finally been resolved, the proof's reception by the mathematical community has been cool.

> . . . mathematicians . . . who were not aware of the developments leading to the proof are rather dismayed by the result because the proof made unprecedented use of computers; the computations of the proof make it longer than has traditionally been considered acceptable.

The length of the proof is due to the huge size of the unavoidable set of reducible configurations—1482 configurations, as compared to the handful (Figure 160) needed for the Five Color Theorem. Having to consider a large number of cases is aesthetically repugnant to a pure mathematician, and if the proof were only unwieldy in the conventional sense, that alone would account for widespread disappointment. But we learn from the next sentence that the proof is more than "unwieldy":

> In fact the correctness of the proof cannot be checked without the aid of a computer.

The proof then is an altogether new kind of proof, bristling with philosophical problems. It cannot be completely comprehended by any human being!

—April 1978

INDEX

SPECIAL SYMBOLS